13.25

Bionomics and Embryology of the
Inland Floodwater Mosquito *Aedes vexans*

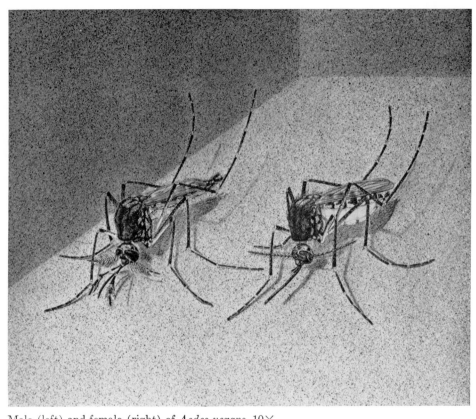

Male (left) and female (right) of *Aedes vexans*. 10×

Bionomics
and Embryology of the
Inland Floodwater Mosquito
Aedes vexans

WILLIAM R. HORSFALL, HARLAND W. FOWLER, JR.,
LOUIS J. MORETTI, and JOSEPH R. LARSEN

University of Illinois Press   URBANA CHICAGO LONDON

CAMROSE LUTHERAN COLLEGE
Library

CONTENTS

ACKNOWLEDGMENTS

This report is based on data and observations accumulated at the University of Illinois over an interval of more than two decades of work with this and other floodwater mosquitoes, as well as related work reported by others elsewhere. Manual and monetary assistance has come from many sources. Assistants to whom much credit is due are George B. Craig, Jr., R. F. Harwood, R. C. Wallace, R. D. Pausch, J. Berger, M. Engelman, W. K. O'Hair, R. Brust, and J. F. Anderson. Personnel and facilities of various mosquito abatement districts in Illinois and of the Tennessee Valley Authority have been generously provided. Eugene Pickard of the latter agency has been particularly helpful. Monetary support has been provided by the National Institutes of Health, Washington, D.C.; Graduate College, University of Illinois; Department of the Army; and the South Cook County Mosquito Abatement District. An additional debt is owed the many whose observations have been referred to in the text. We wish to thank Mrs. Alice Prickett for her hand-drawn illustrations.

Most observations reported from this laboratory were made in Illinois, Alabama, Manitoba, and Michigan or involved populations from these areas maintained in this laboratory.

Publication of this book was supported in part by the U.S. Army Medical Research and Development Command under contract DADA 17-73-C-3003. We also acknowledge support from the U.S. Public Health Service Training Grant USPH GM 1076.

1

BIONOMICS

William R. Horsfall and Harland W. Fowler, Jr.

INTRODUCTION

General

The family Culicidae is well known taxonomically, and much is known about the bionomics of certain domestic species and others essential to maintenance of some of man's most devastating pathogens. Comprehensive bionomic accounts relating to the family are those of Bates (1949), Horsfall (1972), and Clements (1963). The most inclusive account dealing with a single species is that of Christophers (1960) for *Aedes aegypti* (L.). This species is currently popular as a laboratory animal, and it is often a domestic pest and vector of pathogens in domestic and peridomestic foci in and near the tropics. The essential and often widespread association of many members of the family with pathogens causing diseases such as malaria, yellow fever, dengue, filariasis, and encephalitis has stimulated much study of this family. There is a dearth of knowledge about species of mosquitoes that plague man in his various pursuits out of doors. *Aedes vexans* is one that invades urban and domestic areas and ranges over recreational sites. It is also active in propagating causative agents of disease, especially in feral foci.

Nomenclature

Vernacular name: floodwater mosquito, swamp mosquito (Felt, 1904; Smith, 1904). No vernacular name has been approved by the Entomological Society of America, but the name inland floodwater mosquito seems appropriate.

Binomial taxon and synonyms: Aedes *(Aedimorphus) vexans* (Meigen, 1830). (See Stone *et al.*, 1959).

Culex vexans Meigen, 1830
Culex parvus Macquart, 1834
Culex articulatus Rondani, 1872
Culex malariae Grassi, 1898
Culex sylvestris Theobald, 1901
Culex montcalmi Blanchard, 1905
Culicada minuta Theobald, 1907
Culicada eruthrosops Theobald, 1910
Aedes euochrus Howard, Dyar, and Knab, 1917

Trinomial taxa and synonyms:

Aedes vexans vexans (Meigen)
Aedes vexans nocturnus (Theobald, 1913)
  Culicada niger Theobald, 1913
Aedes vexans nipponii (Theobald, 1907)
  Culicada nipponii (Theobald, 1907)

Unless indicated by area or name the trinomial taxon, *A. vexans vexans*, is implied in this report.

Distribution

Geographic. The population as a binomial entity is widely distributed across Holarctica and Pacific Oceania. The most northerly report of occurrence thus far is that of Sommerman (1968), who collected it near Fairbanks, Alaska (N.L. 64°). It is known from Canada south from 55° in Manitoba across the breadth of the land (Hearle, 1932) and all states of conterminous U.S. (called herein USA-48) (Carpenter and LaCasse, 1955) and Hawaii (Joyce and Nakagawa, 1963). It is known on the west side of Africa, eastern Asia, and much of Pacific Oceania from Japan to Samoa. Marshall (1938) considered the species rare in Britain. It is common on flood plains of rivers in France (Gilot, 1969), Germany (Möhrig, 1969), and Czechoslovakia (Trpis, 1962).

References to distribution of this species are abundant. Usually they consist of a listing with no additional information. Some that show the range of the species are included herewith. Adams and Gordon (1943), Annand (1941), Barnes *et al.* (1950), Beadle (1963), Bishopp *et al.* (1933), Blickle (1952), Boddy (1948), Chow (1949), Christensen and Harmston (1944), Darsie *et al.* (1951), Dickinson (1944), Dorer *et al.* (1944), Dyar (1921, 1922), Feemster and Getting (1941), Fellton *et al.* (1950), Ferguson and McNeel (1954), Freeborn and Bohart (1951), Gerhardt (1966), Harmston (1949), Harmston and Rees (1946), Harmston *et al.* (1956), Headlee (1945), Hill (1939), Horsfall (1937), Irwin (1941), Johnson (1959), King *et al.* (1942), Kurashige (1964), Laird (1954), Lake (1967), Love and Smith (1957), MacGregor (1925), McLintock (1944), McLintock and Burton (1967), Mulhern (1936), Myklebust (1966), Niebanck (1957), Owen (1937), Owen and Gerhardt (1957), Post and Munro (1949), Price (1964), Pucat (1965), Quinby *et al.* (1944), Rees and Harmston (1948), Richards (1938), Richards *et al.* (1956), Rowe (1942), Rozeboom (1942), Scanlon and Essah (1965), Schlaifer and Harding (1946), Siverly (1966), Smith (1904a), Stabler (1945), Steward and McWade (1961), Twinn (1953), Venard and Mead (1953), Wilson *et al.* (1946).

Two doctoral theses reporting details of certain aspects of bionomics of this species are those of Fowler (1969) and Thompson (1964).

Reisen *et al.* (1972) reported that the trinomials *A. v. nocturnus* and *A. v. nipponii* occurred on Guam in the Marianas Group.

Ecologic. Aedes vexans vexans is a product of areas where depressions in coarse-textured, alluvial soils are shaded by a herbal or sylvan canopy that are subject to one or more cycles of flooding and drying annually. In such areas the water table rises to inundate the depressions, and then falls below the surface far enough to remove gravitational water from the surface layer of soil. Specifically, the species is a product of the flood plains of streams and rivers; poorly drained, sandy lands where water accumulates as a result of runoff and rain; and margins of impoundments. The North American population is particularly abundant where the continent was left with an uneven topography following the retreat of the last continental glacier. It is now declining over much of this range, as the water table is managed to make the soil suitable for agriculture or urbanization. Incidence of it is increasing along flood plains of rivers wherever runoff is hastened by cultivation and by urbanization, especially where obstruction to flow causes flood plains to be inundated. The European population is largely a product of flooded grasslands (Möhrig, 1969).

Seasonal. In Holarctica this species occurs as eggs at any time of the year, as larvae for a short time whenever developmental sites are flooded by water warm enough to hatch the eggs, and as adults for about five weeks after a brood emerges. Peterson and Smith (1945) found it locally abundant during spring and fall in southern USA-48. Numerous observers have reported it abundant in northern USA-48 and Canada in summer even as late as the time of appearance of first frosts. In any part of this area larvae may be found at different times during the summer, because rainfall causing flooding of larval sites is often variable and local. The relative abundance of adults that follows is usually in proportion to extent of area flooded. Records from abatement districts near Chicago show that adults may be trapped at any time from mid-May until mid-September or even later. Judd (1954) noted that, in Ontario north of Lake Erie, two peaks of adult abundance may occur: one in June and one in early August. Happold (1965) found larvae after mid-May at 54° N.L. in Alberta. Rees (1943) reported that a brood occurs in May, another in June-July, a third in August, and even a partial fourth in September in Utah. Owen (1937) reported that adults usually occur from late spring until September in Minnesota. Callot and Van-Ty (1944) reported that adults may occur in France all summer because several generations are normal. Trpis (1962) noted that along the Danube (48° N.L.) the species is active from May to October.

The occurrence of an early fall brood is variable. Two seasons in the course of two decades of field observations in central Illinois produced wholly dissimilar results. In 1955 some 125+ mm of rain fell on October 5, and all flood plains and depressions were flooded. Larvae were obtained October 6 when up to 269/10 dips were taken. In 1969 similar flooding occurred early in October, and no larvae could be found in any sites. Weather antecedent to the two inundations seemed to be the only difference. Normal seasonal temperatures prevailed in 1955, and lower than average temperatures for a three-week interval prior to flooding occurred in 1969 (Wilson and Horsfall, 1970). During the fall of 1970 a brood appeared as it had in 1955 when warm weather prevailed prior to flooding.

Economic Importance

In late spring, throughout the summer, and even during early fall, this mosquito may be an intermittent pest of man and livestock in North America. Across its range it is the principal cause for daytime annoyance from mosquitoes outdoors, especially in suburban areas. It has been the reason for the establishment of abatement districts in metropolitan areas across central USA-48 and southern Canada.

Regions where it attains maximum annoyance have abundant transient collections of runoff water. Twinn (1931, 1949) said it is the most important rain-pool species in eastern Canada, where it is a transitory pest in urban and rural situations, even in dwellings. It is a major pest along the Atlantic seaboard, especially the northern part (King et al., 1960; Smith, 1904). By all standards it is the most important mosquito in northern USA-48 below the coniferous forest (Owen, 1937; Clarke, 1937). In the northern plains it is a major pest, especially in pastures where certain crops are irrigated (Stage et al., 1952; Tate and Gates, 1944).

Aedes vexans is sporadic and locally severe in south-central USA-48 (Tennessee and Mississippi) (Michener, 1945; Schlaifer and Harding, 1946). In Georgia, where aedines comprise over 97 percent of trapped mosquitoes, A. vexans is locally severe. It is a secondary species in Florida (Knight, 1954). It is reported to be locally severe in southern Europe along the larger rivers. Trpis (1962) considered it the dominant pest along the Danube following spring and summer floods. Adults have appeared in plague proportions along the great rivers of Europe (Möhrig, 1969) and in parts of Russia (Mukanov, 1970).

This mosquito is a significant component in zoonotic reservoirs for Tahyna virus of central Europe and encephalitis viruses of North America and the Orient. It is a good host for Dirofilaria immitis, but its place in the vector complex is not established.

BIONOMICS

Generations

Aedes vexans may have one or more generations a year (Gilot, 1969; Möhrig, 1969; Trpis, 1962), but, because embryos respond erratically to hatching stimuli, generations may overlap in the same site. Any population in the binomial taxon has the capability of a generation about every twenty-five or thirty days when all conditions are favorable. In response to varying combinations of conditions imposed at higher latitudes in Holarctica, generations may be spaced a year or even two years apart.

Accounts in the literature are varied and confusing as to frequency of generations. Part of the confusion results from the fact that generations and broods may become inseparable in the field. Twinn (1931) noted that overlapping generations and erratic hatching within a generation resulted in designating an episode of activity as a brood. Thus he said that broods occur one or more times in Ontario between May and September. Mitchell (1907) reported broods at intervals between February and November in Louisiana and between May and October in New Jersey. McGregor and Eads (1943) reported that a brood may occur at any time of the year in parts of Texas. Owen (1937) stated that one generation is normal each year for a latitude of about 45°

(Minnesota), even though the mosquito is optionally multivoltine (Barr, 1958). Reports by King *et al.* (1960) and Stage *et al.* (1952) show that multiple annual generations do occur. Certainly repetitive generations may occur in areas where impoundments are subject to repetitive changes in water level. Such is probably the case when fresh water spreads over lowland coasts because of tidal conditions.

In areas limited to one annual episode of flooding, only one generation is possible during the year. Others may occur when a site is subject to adequate drying that permits oviposition and completion of embryogeny between each episode of flooding. In the latitude of central Illinois this sequence often occurs as many as three times each summer, especially on flood plains and upland depressions inundated by heavy rain. Because summer rains are erratic in place and amount, a brood may originate from one site at one time and another at a different time. Therefore broods may be numerous and overlap, but generations vary from one to three.

For ten years (1950-59) a record was kept on the times and extent of flooding of a transient slough in Champaign County, Illinois, and the production of broods of *Aedes vexans*. This area was under water for part or all of the winter season during each of the years. When the floods came only during the season of frost (November to March), no broods resulted. One or more episodes of partial or entire flooding came each year between April and October, and each produced a partial or full brood. Five full broods were produced during the ten years when twelve full floods occurred. Nine additional broods came with reflooding of the upper levels following partial drying, and four broods came with partial flooding from the wholly dry state.

Egg

Appearance (Figs. 1-14). Eggs are elongate, fusiform bodies slightly more arched ventrally than dorsally. Newly deposited eggs are veiled in a clear, colorless shroud or pellicle called exochorion (maternal follicular epithelium) that is closely adherent to the contour of the secreted eggshell (chorion). New eggs are wholly encased by the exochorion, while those only a few weeks old may retain little or no pellicle as adherent flakes. Chorion of a newly deposited egg is an off-white, but it darkens rapidly to the definitive color within an hour after deposition at 25°C. The exposed chorion (when viewed against a black background in reflected light from an incandescent source at magnifications of 50× or more) has a bronze hue that permits recognition of the species among contemporary populations in North America (see Horsfall & Craig, 1956).

A mixed population of eggs may vary in length by as much as 200 μ and in diameter by as much as 60 μ. The range in length for 392 eggs showed extremes of variation between 561 and 743 μ. A population from a flood plain in central Illinois (Piatt County) is a bimodal one according to length (Figs. 3, 4). The shorter mode varied between 578 and 693 μ in length, and the long one varied between 677 and 726 μ. Mean length of the former was 658 $\pm$ 2 μ and the latter was 710 $\pm$ 3 μ (Fowler, 1969). Another group of forty-eight eggs taken from the site of a woodland pool in central Illinois ranged in length from 601 to 684 μ (mean 640 $\pm$ 4 μ). All eggs whether long or short were in the range of 165 to 231 μ at greatest diameter. Kliewer (1961) noted that newly deposited eggs of *Aedes aegypti* increased in weight from 5 to 12 μg. Those of *A. vexans* increase in diameter within a day's time after deposition, and within a few days they assume their definitive shape.

Surface of an egg is marked with a distinctive pattern of polygons (Figs. 6, 9, 12, 14). Each polygon is about 1/20 the length of the egg, and its disc is a flattened area bounded by lace-like lines when seen at magnifications of 400$\times$ or more (Figs. 12, 14). Furthermore, the polygons are elongated in line with the long axis of the egg so that at low magnifications (10 to 20$\times$) an egg appears to be striped longitudinally. The reticulation, color, shape, and size of eggs provide diagnostic characters for recognition at the binomial level (Ross and Horsfall, 1965; Craig, 1956; Horsfall *et al.*, 1970). The distinctive features of the reticulation may be seen on whole eggs viewed at magnifications over 60$\times$ in reflected light against a nonreflective black background. Minute details of the sculpturing may be seen when sheets of chorion are partially cleared (Craig, 1955) and phase-contrast microscopy at magnifications of 400$\times$ is used. More recently the scanning electron microscope has provided visualization of all markings at any level of magnification from 100 to 2,500$\times$ or more (Figs. 1-14). Photographs at 100$\times$ with this device provide details of size, shape, and marking as they appear in reflected light of the microscope.

The scanning electron microscope reveals minute differences that are diagnostic for populations within the binomial taxon. Figures 1-14 showing the reticulation of populations from Florence (northern), Alabama, Mermet (southern), Illinois, and Piatt County (east central), Illinois, illustrate some of the differences. The reticulation along the center of an egg from Alabama has a smooth disc in each polygon, and the lace-like margins of each polygon are wider. The population from Mermet, Illinois, has a rugose disc, as does the one from Champaign County, Illinois (Figs. 6, 13, 14). The short form from Piatt County has a rumpled surface on the disc, while the long form has a nearly smooth one. Figures showing the micropylar ends amplify the differences (Figs. 10, 11, 13).

Focal distribution. Eggs are found on soils (Dupree, 1905; Bodman and Gannon, 1950) subject to transient flooding. They are on soil surfaces even when those surfaces extend below ground level as cracks, crayfish holes, and hoof prints. Generally, the situations involved are shallow depressions in field, forest, and flood plain where water from rain and runoff collected a few days or a few weeks earlier in the season. Abundance of eggs in any site varies in proportion to the length of time the soil surface has a moisture level attractive to ovipositing adults. Bare, uncracked, and unshaded soil seldom contains eggs. Cracking encourages oviposition on bare soil. Canopy shade, especially dense, low, herbal shade, also encourages oviposition. Loose layers of herbal detritus as soil cover, especially when shaded, indicate large numbers of eggs. All of these phenological details reflect retention of moisture at levels attractive to ovipositing females over a wide range of time.

In order to appraise the incidence of eggs, a census procedure yielding a consistently high percentage of eggs is required (Horsfall, 1956a). In essence the procedure providing data reported herein involved sieving, flotation, and visual separation in series. Samples composed of surface soil and overburden of detritus over an area of 225 cm² and 2.5 cm deep were considered representative. They were obtained by inscribing the area beneath a template with a sharp mason's trowel (Fig. 15), then cutting the sample free and placing each sample in a bag for transfer to the laboratory. (For sampling by transect, see Lefkovitch and Brust, 1968.) Each was screened by washing through sieves with meshes of 4, 8, 16, and 40 squares to the linear inch for removal of coarse herbal detritus. Eggs, sand, and minute detritus were caught on a screen with mesh of 100 holes to the linear inch while silt passed through with the water. Eggs and debris of similar density were freed from the heavier mineral components by mixing the residue on the 100-mesh screen with a saturated solution of salt (NaCl) in water. Eggs floated, while sand and heavy debris settled to the bottom. Eggs together with other residue were screened from the salt water and submerged in tap water. Final separation from accompanying debris was achieved by pipetting eggs to a dish of water. This was done while viewing at a magnification of 10-20× with a stereomicroscope.

A system for recognizing and separating eggs to species was devised incident to the survey (Horsfall & Craig, 1956). The order of dependability of the separation process is shown by records of five consecutive samples each containing 100 eggs from which 81, 82, 86, 87, and 89 eggs respectively were recovered.

Aedes vexans is in the ecological category of floodwater mosquitoes because the eggs are present on soil at the time they are inundated by rain or by runoff water trapped as pools on the ground. Eggs begin to appear in sites as the water recedes following inundation in spring and summer (Horsfall, 1963; Wilson and Horsfall, 1970). They appear first on the soil that is moist but not water-logged, which means they are placed in a zone above the water table at the time of deposition. When summer rains moisten but do not flood oviposition sites, eggs may be deposited anywhere over a depression where it is suitably moist.

The general aspects of distribution of eggs is shown in Table 1. A woodland depression some 660 m in length by an average of 15 m wide was sampled in November along profiles 15 m apart. Across the width at each profile five samples representing marginal, intermediate, and low horizons were taken. A total of 215 samples (of 225 cm²) yielded 34,765 eggs (450/625 cm²) of *Aedes vexans* alone. The number of eggs in the combined five samples at each profile varied from a low of 101 eggs to a high of 3,964 (Table 1).

The distribution of eggs as affected by horizon, runoff, and surface detritus was determined in two woodland depressions in Champaign County, Illinois (Tables 1 & 2). Both sites were elongate embayments of the former channel of a relocated stream under full canopy shade with a border of elm and willow. All water in the embayments accumulated as runoff from a watershed of about twenty acres west of them. Site A was a pool about 36 m long by 12-15 m wide in normal flood; site B was slightly longer and, on the average, wider than A. Both were subject to flooding above the normal level for a few feet on either side into a surcharge zone. Site A was the deeper depression by 15 cm or more and had much more bottom area below the 60 cm contour.

Site A had three areas where the surface moisture was retained longer than elsewhere. The entire west side was bounded by a high (2 m) bank of packed sandy soil which held water that slowly percolated downward and moistened the bottom along the west side of the depression. Profile 1 crossed the site over an area where prolonged high moisture levels prevailed in the soil during the summer. Profile 4 crossed a silted delta moistened from surface water of a runoff channel that entered onto the profile. The east bank was flat and very like that of site B; however, the bottom was seldom moist.

TABLE 1

Distribution of Eggs of Floodwater Mosquitoes in an Upland Woodland Slough Based on Five Samples (=area: 1125 cm²) Taken on Profiles at Intervals of 15 m in Champaign County, Illinois, during Late Fall

Profile	Number of eggs								
	Aedes vexans	Aedes trivittatus	Aedes stimulans	Aedes canadensis	Psorophora ciliata	Psorophora horrida	Aedes dupreei	Aedes sticticus	Unidentified
0	413	232	20			18			
1	844	161	90			82			
2	760	216	18	5		67			
3	921	291	23		1	69			1
4	840	164	37		1	54			3
5	421	46	8			3			
6	530	15	9			23			2
7	1,084	7	37			1			1
8	549	42	1			9			
9	1,040	60	19			37			3
10	734	101	146	5		6			
11	272	2	4	1	1	22			
12	298	72	25	1		4			
13	354	86	24	15		58			
14	570	90	25	2		73			
15	319		43			162			
16	803	21	5			5			
17	975	161	40		1	110			
18	522	92	66			39			
19	493	82	18			31			
20	3,964	51	6			52			
21	1,738	280	68			76			

TABLE 1 — Continued

Profile	Number of eggs								
	Aedes vexans	*Aedes trivittatus*	*Aedes stimulans*	*Aedes canadensis*	*Psorophora ciliata*	*Psorophora horrida*	*Aedes dupreei*	*Aedes sticticus*	Unidentified
22	693	95	34		4	102			
23	860	25	38			36			
24	694	42	1						
25	408	13	16			3	4		
26	294	1	15						
27	743	6	8				9		
28	380	7	2			5	12		
29	1,065	81	35			74			
30	1,490	58	49			7			
31	887	44	5			30			
32	611	3	33					3	
33	730	47	78						
34	158	1						9	1
35	101	3							1
36	137	76	115			3			
37	225	60	29			25			
38	808	42	6			3			
39	2,849	68							
40	155								
41	126	1	1						
42	949	7	14						
43	1,957								
Total	34,765	2,952	1,211	29	8	1,289	25	12	12

Site B had one area (crossed by profile 4) where surface moisture persisted longer than elsewhere because water from a runoff stream entered across a silted delta after summer rains. The area was moist enough to support a dense stand of lizard's tail (*Saururus*). Both east and west banks were low, and the bottom was usually dry.

Serial samples (of 225 cm²) were taken along each of four profiles across each embayment so as to form continuous bands 15 cm wide extending from the normal flood line on one side to the same horizon on the other side. Since the samples adjoined each other, the composite at each profile was a band 15 cm wide and 10 m or more long. All samples were taken after oviposition had ceased and before the soil was frozen (November and December).

History of the water table for the summer prior to sampling showed five changes in water level. Adults of the early brood were permitted access to horizons progressively exposed for oviposition from the surcharge zone to 32 cm below it between April 29 and May 26. Water rose to a level of 22 cm below flood May 27 and had disappeared from the site by June 13. The site was partially submerged again June 19 and was dry by July 2. The water rose gradually to 25 cm between July 8 and 15. A rain in the interval July 31-August 1 brought the water level up to 22 cm. The final drying was complete by August 18. The wet zone was most continuous below 30 cm, but some of this area was subjected to alternate submersion and drying. Profiles 1 and 4 of site A and profile 4 of site B were across regions where input of water by runoff kept the soil moist longer than elsewhere in the embayments. Opportunities were thereby provided for eggs to hatch and for others to be deposited by successive generations. By the time of the survey, the population of eggs represented one or more generations of each of the species present.

In both sites eggs were placed at all levels below normal pool level for the season (Tables 2, 3). The total number of eggs and the number per unit area were greatest below the 30 cm contour. Actually, the contour significantly affected distribution only insofar as the slope affected retention of the moisture that attracted the ovipositing females. Profile 4 across the south end of site B was kept moist up to the 15 cm contour because of a dense, low canopy of lizard's tail, and the number of eggs per unit area was higher than in any area on the other profiles. The area of profile 1 in site A bore the greatest number of eggs and the most eggs per unit area. It was covered by an absorbent layer of herbal detritus and many branches that had dropped from the canopy. It, too, received runoff water from the surrounding area and remained moist most of the summer.

TABLE 2

DISTRIBUTION OF EGGS OF *Aedes vexans* ALONG PROFILES ACROSS ALL HORIZONS OF TWO WOODLAND SITES DURING THE WINTER, CHAMPAIGN COUNTY, ILLINOIS

Number of eggs per 625 cm²

Horizon	Site A				Site B			
	Profile 1	Profile 2	Profile 3	Profile 4	Profile 1	Profile 2	Profile 3	Profile 4
0 to −15 cm east	10	1	3	14	9	83	83	5
−15 to −30 cm east	185	1	95	0	92	134	90	790
−30 to −45 cm east	2,248	174	575	617	0	324	582	780
center (−45 to −75 cm)	1,112	292	262	334	84	127	71	432
−30 to −45 cm west	1,228	0	33	47	3	158	84	948
−15 to −30 cm west	162	0	0	10	277	200	10	890
0 to −15 cm west	0	0	0	2	15	3	0	20

TABLE 3

DISTRIBUTION OF EGGS OF *Aedes vexans* IN TWO WOODLAND SITES,
CHAMPAIGN COUNTY, ILLINOIS

Site	Depth below flood line (cm)	Area sampled (cm²)	Eggs present		Percent
			Total	1,000 cm²	
	0 to −15	6,974	35	5	<1
A	−15 to −30	4,720	591	125	1
	−30 to −45	11,520	57,364	4,979	98+
	0 to −15	8,350	458	55	1
B	−15 to −30	7,497	3,851	515	10
	−30 to −45	77,050	34,436	447	89
Total		116,111	96,735	833	..

While it is true that most eggs (as totals and per unit area) were below the 30 cm contour, it is also true that this region had the greatest range in numbers per sample. The variable was of the degree of exposure of the soil to drying. Where the soil absorbed and held water because of detritus and shading by low herbal canopy, numbers were maximal. Where absence of these factors permitted rapid drying of the soil, few or no eggs were found, as was the case near the top of pool level.

A series of samples taken in a woodland depression late in the summer indicated variation in specific features of oviposition sites. The walls of crayfish holes were often abundantly populated with eggs, especially when entrances to the holes were well below the line of maximum flood. Yields from four holes were 190, 315, 584, and 421, when holes were near the bottom of the depression. Only one egg was found in a hole where the entrance was above the flood line. Wide cracks in silt in the bed of a large lake that had been drained yielded 120, 39, 453, and 100 eggs from an area of 112 cm² of soil surface from the faces of each of four cracks. Eggs were found along the walls of cracks 45 cm below the surface. Samples taken from surface areas of the depression covered by herbal detritus yielded comparable populations per unit area. Tree branches and logs lying on a woodland depression encourage oviposition on underlying soil more than do areas covered by leafy detritus. For another year and under other conditions, comparisons between cracked soil and herbal detritus as sites for eggs showed the following relative incidence: wall of crack under dense, smartweed canopy in a drained slough: 300+/625 cm²; soil above cracks: $100+/625$ cm^2; bunch grass base: $3,200+/625$ cm^2; detritus beneath buttonbush: $3,100+/625$ cm^2.

The coincidence of shade and the nature of detritus on incidence of eggs is shown by a series of samples taken about the time of maximum oviposition. Seven samples (1,575 cm^2) were taken from soil overlaid with herbal detritus beneath a dense canopy of lizard's tail 1 m high; three samples (675 cm^2) came from a site similarly shaded without detritus; four samples (900 cm^2) were from a site covered by woody-leafy detritus under high willow canopy; and five samples (1,125 cm^2) were from soil beneath woody-leafy detritus without canopy. The area of lizard's tail canopy over herbal detritus yielded $2,500+$ eggs/625 cm^2; another covered with detritus and shaded by willows yielded $1,000+$ eggs/625 cm^2; the area with lizard's tail without detritus yielded $100+$ eggs/625 cm^2, and the unshaded area with detritus yielded $400+$ eggs/625 cm^2.

Eggs on flood plains of rivers generally were absent from zones where water flowed unimpeded across the soil. When structures, detritus, vegetation, or natural obstacles inhibited scouring, eggs were found. A series of samples (total of 1,575 cm^2) taken from the flood plain of the Sangamon River at a site covered with woody detritus and subject to the sweep of the current during times of flood yielded no eggs. Topsoil in such situations was coated with a fine silt that often cracked and curled when dried.

Wherever highways and rural roads cross a flood plain on a land fill which obstructs the free flow of the river in flood, eggs will be found on the flood plain either above or below the fill wherever the force of the current is broken. At one such site on the Sangamon River in Piatt County, Illinois, eggs were found on the flood plain on both sides of a land fill. Near the land fill and 100 m away from the channel, 111 eggs per 625 cm^2 were found near the fill in a narrow zone upstream from the obstruction. On the downstream side of the same obstruction, eggs were found everywhere the current had been deflected. About 100 m from the channel and within 30 m of the fill, eggs were very abundant in a zone of woody detritus (1,566 eggs/625 cm^2). Even in zones of minimal detritus, the numbers were about the same below the obstruction as they were upstream in seemingly attractive oviposition sites.

The incidence of eggs on a river flood plain varied widely on the downstream side of a highway fill. Nine transects were examined after oviposition had ceased in late fall. The area chosen (in Piatt County, Illinois) was subject to transient overflow but free of current and was well shaded by a canopy of trees in summer. Transects were within an area 45 x 45 m. The surface soil along strips 15 cm wide and 60-300 cm long was removed in segments 15 x 15 cm square and processed for eggs. Four of the strips were covered with broken twigs and herbal debris. Five of them were overlaid by large logs or were beneath piles of criss-crossed logs that had drifted together. The areas covered by broken woody debris had from 348 to 910 eggs per 625 cm^2, according to the depth and kind of detritus present (Table 4). The incidence of eggs varied from 980 to 3,220 eggs per 625 cm^2 beneath piles of drift-wood, where fragments of bark and twigs were abundant.

The Des Plaines River in Cook County, Illinois, is a sluggish stream carrying a heavy load of organic material in suspension. It crosses a sandy plain for much of its course. In part it traverses wooded areas, and some of the flood plain is grassy or covered by herbaceous canopy. Many obstructions are present as land fills for railroads, roads, and streets. A grassy area downstream from a highway land fill was sampled along a profile from the horizon of the most recent flood to the zone of water-logging caused by partial rises of the stream. The surface was largely covered by grass with patches of smartweed (*Polygonum*) and *Juncus*. A layer of detritus composed of bark and herbaceous stems covered much of the area. The transect sampled was about 45 m long and extended from the flood line downward to a level about 45 cm below that of maximum flood for the year. The horizon from 0 to 15 cm below maximum flood had 63 eggs/625 cm^2; that 15-30 cm below had 91 eggs, and that 30-45 cm below had 214 eggs/625 cm^2. Evidently the lower horizon was moist longer or more often than upper ones.

The effect of wet soil on incidence of eggs is further illustrated on a vast series of grassy swales south of Lake Michigan in Cook County, Illinois. Water becomes trapped between ridges after each period of heavy rainfall. The area is covered by a dense canopy of grass (*Spartina*) 60-90 cm high. Samples were taken at intervals along a profile across one depression at 3 cm contours from a zone near normal high water downward across the grassy zone to a *Juncus* zone above muck. Table 5 shows the incidence of eggs in early July before the area was entirely flooded on July 15, during the water-logged phase (July 30 and August 13), and after the season for oviposition had passed (October 11). The eggs per 625 cm^2 were over 1,000 in the horizon at or near the normal high water mark before the flood and again at the end of the season. During the time the soil was wet, the number present in this zone was about 1/20 of that earlier in the season and at the end. The number per 625 cm^2 in the zone 9 cm below high water was 1/3 to 1/5 that higher up, but the area involved bore more eggs than did the higher zone because of the greater width. The horizon −12 cm to −21 cm (i.e., level below maximum flood) had about 1/20 the

TABLE 4

INCIDENCE OF EGGS OF *Aedes vexans* ON SOIL ALONG CONTINUOUS STRIPS 15 cm WIDE DOWNSTREAM FROM A HIGHWAY FILL ON THE FLOOD PLAIN OF THE SANGAMON RIVER, PIATT COUNTY, ILLINOIS, IN LATE FALL

Profile strip	Depth below normal flood (cm)	Area sampled (cm²)	Number of eggs/625 cm²				
			A. vexans	A. trivittatus	Other Aedes	Psorophora ferox	Psorophora ciliata
Detritus	−15 to −30	1,730	374	16	0	36	0
Detritus	0 to −30	4,750	580	16	1	50	1
Large log	0 to −30	5,150	910	48	15	132	0
Detritus	0 to −30	7,900	348	33	9	26	1
Logs	−15 to −30	3,150	1,312	190	2	152	5
Logs	−15 to −30	2,700	980	238	22	59	1
Log jam	−15 to −30	1,120	2,360	355	20	160	2
Log jam	−15 to −30	900	1,708	130	2	132	5
Log jam	−30 to −45	1,340	3,220	147	2	24	6

TABLE 5

INCIDENCE OF EGGS OF *Aedes vexans* IN A GRASSY SWALE BETWEEN TWO BEACH
LINES LEFT BY RECESSION OF LAKE CHICAGO TO BECOME LAKE MICHIGAN
IN COOK COUNTY, ILLINOIS

Horizon below flood (cm)	Number of eggs/625 cm²			
	Pre-flood	Post-flood		Dry
	July 1-8	July 30	August 13	October 11
Normal flood line	1,080	52	55	1,180
3 to 9	165	24	71	364
12 to 21	49	Flooded	Flooded	48
24 to 42	12	Flooded	Flooded	0

number of eggs per 625 cm² of that in the area of the highest horizon. No eggs were laid in July and August because the area was flooded. The horizon −24 to −42 cm bore almost no eggs early in the season and was flooded so late that none was deposited before the season was over.

Extensive cattail (*Typha*) marshes in Cook County, Illinois, having contour difference of less than one-half meter across hundreds of acres are sites for vast numbers of eggs. The soil is a soft black loam very high in comminuted herbal remains above water-logged sand. *Aedes vexans* was the only species present in several hundred samples examined.

One such site of some 400 acres in an extinct channel adjacent to the Calumet River yielded 874 eggs from 20 samples (4,500 cm²) along a line of about 300 m across the marsh. This represents 121 eggs/625 cm². While the yield per sample is low, areas available for oviposition are vast, thereby providing major sources of this mosquito. Sites such as this are subject to exposure above water or to inundation according to fluctuations in the level of Lake Michigan.

In Michigan, eggs have been found in woodland depressions in sandy soils where the canopy is deciduous (commonly maple) trees. Eggs were absent where the canopy is spruce and the pools of coffee-colored water are lined with sphagnum. No eggs have been found on flood plains where the watershed has a heavy growth of spruce. Eggs may be very abundant in naturally impounded areas adjacent to lakes in southern Michigan, especially when shaded by maple canopy. Large areas yielding over 400 eggs/625 cm² have been found in the region from the latitude of Flint (near 43°) and southward. From 44° northward the species is local and sites are rare. Eggs have been found along the central ridge, where the soil is sandy and devoid of overburden of silt or muck. No densities comparable to those of the southern part of the state have been found.

Vertical distribution. Filsinger (1941) found that eggs of a population in eastern USA-48 were on or below the soil surface. They were most numerous in the clumps of grass at the ground line or under detritus on the surface. None was as much as 50 mm above the surface. It is of interest that about 40 percent of the eggs that survived the winter were 25-50 mm below the surface in June. In August many were in the layer 25 mm below the soil surface, but none was 50 mm below. In the Pacific Northwest Gjullin *et al.* (1950) found eggs in moss growing on trees, logs, and concrete abutments, but most eggs were on soil beneath herbal detritus. In Illinois all eggs are in contact with soil.

Horsfall (1963) noted that eggs of *Aedes vexans* are deposited on the surface of soil and usually beneath herbal detritus. All do not remain on the surface even in uncracked soil. Late in October a sample of soil from an oviposition site was taken to a depth of 7.5 cm and examined in layers of 5 mm. Samples were taken by forcing metal tubes into the soil to a depth of 7.5 cm after removal of the overburden of detritus. The column of soil in each tube was forced out with a plunger in increments of 5 mm. Eggs were removed in the manner described above for census procedure. A total of 34 eggs was recovered from beneath the surface. Thirty percent were in the upper 5 mm. Eighty-five percent were found in the top 2.5 mm, and 15 percent were below this. Most of these eggs were viable.

Eggs in some situations are covered by more and more soil as the season advances. During 1969 Wilson and Horsfall (1970) reported progressive burial of eggs in an embayment on the flood plain of the Sangamon River, Piatt County, Illinois. The area was shaded by a heavy stand of canary grass (*Phalaris*). The soil is silt over sand and is subject to flooding when the river covers its entire flood plain. During 1969 the area was flooded in April, causing overwintering eggs to hatch. Adults had emerged by the end of April. The area was not inundated again that season. Residual eggs after the April flood numbered 78/625 cm² in the top 25 mm of soil over the area. The first new eggs appeared June 5, and few new eggs were deposited after July 1 that year. Table 6 shows the vertical distribution of eggs in the interval July 2 to November 4, a period after deposition had ceased and before the soil became frozen. All of the eggs were at the surface in early July and would have hatched had they been flooded, as was shown by laboratory

TABLE 6

VERTICAL DISPLACEMENT OF EGGS OF *Aedes vexans* IN SOIL ON A FLOOD PLAIN
IN CENTRAL ILLINOIS AFTER CESSATION OF OVIPOSITION

Depth (mm)	Percentage eggs/625 cm²					
	July 2	July 15	Aug. 2	Sept. 6	Sept. 22	Nov. 4
0–5	88	6	47	31	31	25
6–10	0	10	25	26	31	20
11–15	5[a]	8	7	13	13	20
16–20	5[a]	6	7	10	13	4
21–25	0	6	7	6	6	9
26–30	0	2	7	7	4	6
31–35	2[a]	0	0	3	2	9
36–40	0	0	0	4	0	6
41–45	0	0	0	0	0	1
46–50	0	0	0	0	0	0

[a] Probably residual eggs from previous season.

tests. Over two-thirds were still at the surface by mid-July, less than half were on the surface by early August, and only one-fourth were at the surface by wintertime. In late fall some eggs had become displaced downward as far as 45 mm, 16 percent were more than 30 mm deep and over half were over 10 mm in the soil. The buried eggs retained viability but could not hatch *in situ*.

Seasonal occurrence. Eggs may be found during any month of the year. A site in central Illinois was flooded in October, 1955, and remained under water during the winter and spring of 1956. A sample of soil (225 cm²) bore 285 eggs (70% viable). No spring hatching occurred at this site. By early June the site had dried, and June 11 samples yielded 605 eggs; 11 (2%) were 1956 eggs. Samples taken June 14 yielded 1,802 eggs; 514 (31%) were new. More new eggs were deposited following a soaking rain June 19. Another soaking rain July 8 induced more oviposition. Before the eggs deposited in July were embryonated, sites bearing some were flooded on July 15. A sample taken under water July 18 yielded 229 eggs (all new). A partial flood in August showed no eggs below the water, 347 new eggs/625 cm² at the water line, 152 new eggs/625 cm² in the moist zone, and 0 eggs/625 cm² in the dry zone.

In 1954 a few residual eggs (7-10/625 cm²) survived (see vertical distribution) and new (i.e., 1954) eggs began to appear May 10 and were being deposited at peak numbers by May 21. No viable residual eggs were found June 4. After the partial floods in June and July new eggs appeared in the wet zone as the water receded. No new eggs appeared after July 30.

Associated species. Aedes vexans is likely to occur as eggs along with other species of floodwater mosquitoes over most of its range. Woodland depressions of central Illinois, where *A. vexans* is by far the dominant floodwater species (50-90%), usually contain *A. trivittatus* (Coq.), *A. stimulans* (Walker), *A. sticticus* (Meigen), or *Psorophora horrida* (D. & K.) (Table 7). Additionally, *A. canadensis* (Theo.), *cinereus* (Coq.), *dupreei* (Coq.), *Psorophora ciliata* (Fabr.), *P. howardii* (Coq.), *P. ferox* (Humb.), and *P. varipes* (Coq.) are variably abundant. *Aedes vexans* tends to be found in direct association with all of the species listed above, but the incidence of one or the other species may be greater according to horizon, soil, detritus, and season.

Aedes vexans is dominated by *A. canadensis* in some sites in southern Illinois. This is particularly true wherever a soil is clay, where the depression is overlaid with leaves of oak and a few other deciduous trees, where the canopy shade is entire in summer, and where the understory is light or lacking. A series of samples showed the latter present at a rate of $139/625$ cm^2, while the former was present at a rate of $11/625$ cm^2. *Psorophora ferox* was second in abundance with 80 eggs/625 cm^2. *A. sticticus, trivittatus, dupreei,* and *P. varipes* were also present at rates of $1\text{-}2/625$ cm^2.

The soil from hollow butts of tupelo gum growing on the flood plain of a sluggish bayou (Cache) in extreme southern Illinois yielded eggs of *Aedes vexans.* Larvae of *Aedes thibaulti* D. & K. are found in numbers in such sites, but eggs have never been taken with those of *A. vexans.*

Depressions bearing deposits of clippings from manufacture of composition shingles (roofing felt covered with ceramic grit) yielded mixtures of *Aedes dorsalis* (Meigen) and *vexans* in Cook County, Illinois. Six samples (1,350 cm^2) yielded 847 of the former and 1,417 of the latter in northern Illinois. The former species was the only one present in a nearby area where the soil was overlaid by detritus from a brickyard and fly ash from smokestacks. The surrounding area is dominated otherwise by *A. vexans.*

Aedes vexans does occur in conjunction with *sollicitans* (Walker) in areas of southern Illinois subject to inundation by effluents from plants washing newly mined coal and by seepage from piles of rocky waste ("gob") removed from coal mines. Usually the former species occurs in the less acid marginal areas. In one instance a sample (225 cm^2) on a horizon near the margin of a depression shaded by barnyard grass (*Echinochloa* sp.) yielded 21 *A. sollicitans,* 55 *vexans,* 1 *trivittatus,*

TABLE 7

INCIDENCE OF EGGS OF FLOODWATER MOSQUITOES ALONG EIGHT PROFILES IN A WOODLAND DEPRESSION LESS THAN 30" (75 cm) DEEP AND ABOUT 50' (15 m) WIDE AND 200' (60 m) LONG, AFTER LEAF DROP, CHAMPAIGN COUNTY, ILLINOIS

Profile	Number of samples	Number of eggs									
		A. vexans	A. tri-vittatus	A. sti-mulans	A. stic-ticus	A. cana-densis	Psorophora horrida	P. ferox	P. ciliata	Total	A. vexans %
1	94	32,043	941	10	0	1	58	0	29	33,082	97
2	83	7,026	11	16	0	21	18	21	0	7,113	97
3	103	7,799	203	2	0	14	4	0	0	8,022	97
4	114	11,090	216	1	2	8	67	3	0	11,387	98
5	131	1,888	353	762	0	23	156	3	0	3,185	60
6	136	5,746	1,219	269	38	23	771	22	2	8,090	70
7	136	3,954	276	197	25	43	1,008	65	16	5,584	69
8	117	27,487	3,141	207	15	47	1,943	211	25	33,076	83
Total	914	97,033	6,360	1,464	80	180	4,025	325	72	109,540	89

and 20 *canadensis*. A pile of brush in the same horizon yielded 6 *A. sollicitans*, 134 *vexans,* and 16 *Psorophora ferox*. A wide delta covered with very fine silt bore a pure stand of *A. sollicitans* in vast numbers (10,455 eggs in one sample, for instance). Areas marginal for *A. sollicitans* bore *A. vexans* and *Psorophora confinnis* (L.-A.). Wherever *A. sollicitans* and *A. vexans* were coexistent, the soil was coarser and largely devoid of the red deposit characteristic of sites of high incidence of the former, or shaded by grasses or trees where the acid water from coal washers was diluted by runoff from the watershed.

Aedes vexans is common and even abundant on the sandy flood plain of the Tennessee River downstream from the land fill leading to the bridge at Florence, Alabama, where the area is largely overgrown by pine and deciduous trees. A collection in the winter of 1968 showed that *A. vexans* occurred with *A. atlanticus* D. & K. and a few woodland *Psorophora*, notably *P. varipes*. This same area was sampled in 1956, when the area was largely an open pasture with few scattered trees. At that time *Psorophora* species dominated the population by far, with *A. vexans* rare. A sample (225 cm²) in *Juncus*, for example, yielded 224 *P. confinnis*, 206 *P. cyanescens* (Coq.), 116 *P. discolor* (Coq.), 3 *P. ciliata*, 3 *A. vexans*. Another sample at the base of a young ash tree yielded 313 *P. cyanescens*, 94 *P. confinnis*, 26 *P. discolor*, and 71 *A. vexans*. Other samples were similarly dominated by *Psorophora* species. Within twelve years the relation between *A. vexans* and *Psorophora* had been reversed.

In a shallow depression bordered by small cypress trees with a fringe of a *Cornus* shrubbery in western Tennessee (runoff area above Pickwick Reservoir), *A. vexans* was found with *atlanticus, dupreei, trivittatus, Psorophora ferox,* and *P. varipes*. In 1956 *A. atlanticus* was the dominant species. In 1968 *A. vexans* was by far the dominant one.

Eggs of *Aedes vexans* in southern Michigan have been found in association with those of *stimulans, sticticus, cinereus,* and *trivittatus*. Wherever the black-legged *Aedes* such as *Aedes punctor* (Kirby), *communis* (DeGeer), and *trichurus* (Dyar) are abundant, *A. vexans* is wholly absent or very local in distribution in Michigan.

Development of embryo. Embryos of *Aedes vexans* begin development as soon as eggs are deposited, and they are structurally mature within a few days. Regional populations may vary in their rate of maturation, but no delayed or suspended morphogenesis is known. Dupree (1905) noted that eggs in natural sites were embryonated within five days in Louisiana. Ten days is minimal pre-hatching time for a California population (Jones and Arnold, 1952). Embryos of a population in central Illinois were hatchable within six to ten days at 25°. Gjullin *et al.* (1950), working with a population in the Pacific Northwest (USA-48), found embryogeny to be erratic (9% completed it in 4-6 days, and 75% completed it in 8-10 days). Möhrig (1969) reported four to eight days for embryogeny of a European population. Embryos of an Illinois population matured when incubated at all temperatures between 10° and 30° (Table 8). Using sclerotization of the hatching spine as evidence of a mature embryo, about twenty-four days (564 hours) were required for 50 percent of embryos to mature at 10°, seven days at 18°, four days at 25°, and about three days at 30°. No embryos survive at constant 40°. Temperatures of 30° occur at times in woodland oviposition sites in central Illinois; no adverse effects on embryogeny have been noted.

Longevity of embryo. Embryos may live a year or more, and an idea is prevalent that survival for many years is possible. They have survived in this laboratory for two years in a moist chamber at 4°. Eggs of a population in central Illinois have survived outdoors for nearly two years when kept free from natural hatching stimuli.

Reports on durability of eggs based on field collections or observations indicate a wide range in time. Miller (1930) concluded that most eggs that survive the winter hatch during the summer after deposition. Strong (1940) in a summary report stated that eggs remain viable at least six years. Annand (1941) in a similar report quoted a figure of seven years, with less than 1% viable after five years. Gjullin *et al.* (1950) caged a naturally infested area in the Pacific Northwest and found that embryos survived well through the first two summers, and a mortality of about 70-80 percent occurred in the third year. A small fraction of the population survived into the fifth year. Breeland and

TABLE 8

Interval Required for Embryonation of 50% or More of Eggs of
Aedes vexans at Static Temperatures

Temperature ± 0.5°	Number exposed	Embryogeny completed (hours)
10	45	564
18	58	166
23	60	101
25	60	96
30	60	66
40	60	0

Pickard (1967) reported that a wild population in cages on a flood plain survived into the fourth year. Obviously embryos have a capability of prolonged survival, but the significance of this fact on maintenance of a field population depends much on their specific location and frequency of inundation (see *Eggs: Focal distribution, Vertical distribution*). Eggs of an Illinois population kept at 25° in the laboratory and exposed to hatching stimuli at weekly intervals were not hatchable after forty-six weeks (Table 14). For the first twenty-eight weeks hatchability was 93-95 percent. Between the twenty-ninth and forty-second weeks a steady decline in viability was noted to an average of 65 percent for the interval. After forty-two weeks embryos responded poorly (18%), and after forty-seven weeks none hatched. Eggs on the surface and subject to inundation are unlikely to resist hatching during the second summer.

Latency of embryo. Embryos have the faculty of becoming dormant while retaining a latent capability for hatching, or they may be induced to hatch without delay. Dyar (1902) referred to a temporary latent possibility when he found larvae immediately after flooding of a site where eggs must have been present before flooding. Dryness — that is, the absence of water deep enough to flood eggs — is one cause for latency of embryos (Eckstein, 1919). Latency may persist, however, even when eggs are submerged. Mail (1934) kept a series of eggs without their hatching for an interval of eighty-six days in the fall of the year. Undoubtedly latency is more or less prolonged according to season or environment in which eggs were deposited. Gjullin *et al.* (1950) found that eggs deposited in the fall might not be induced to hatch even after four weeks of exposure to 21°. Dyar (1902) noted that eggs of a New Hampshire population hatched erratically when brought to the laboratory in September.

Eggs of an Illinois population brought into the laboratory from the field in late November after being subjected to transient freezing temperatures but before prolonged freezing of the ground required preconditioning of several days before they responded well to hatching stimuli. No larvae hatched from 360 eggs so exposed after two days at 25°. After seven days, 20 percent of 542 eggs hatched. After fifteen days, 90 percent of 318 eggs hatched.

Induction of latency by exposure to various thermal regimens varied according to the degree and duration of exposure (Table 9). As shown, embryogeny was completed at 25°; then eggs were stored at different temperatures in a moist chamber for a total of twenty-five additional days. Latency was induced at all temperatures below 25°. Twenty-eight days after storage at 23°, 80 percent of the embryos remained

TABLE 9

EFFECT OF EXPOSURE FOR 25 DAYS TO SIX DIFFERENT CONSTANT TEMPERATURES ON HATCHING OF EGG LOTS CONTAINING 50 EMBRYOS OF *Aedes vexans* PREVIOUSLY CONDITIONED BY STORAGE AT 25° FOR 25 DAYS AFTER OVIPOSITION[a]

Holding temperature ± 0.5°	Eggs hatched									
	0 days after storage		7 days after storage		14 days after storage		21 days after storage		28 days after storage	
	No.	%	No.	%	No.	%	No.	%	No.	%
4	0	0	37	74	44	88	36	72	36	72
10	0	0	47	94	16	32	12	24	3	6
18	0	0	24	48	10	20	0	0	0	0
23	0	0	50	100	41	82	44	88	10	20
25	48	96	50	100	50	100	50	100	50	100
30	.	.	49	98	44	88	50	100	50	100

a All hatching at 25° in nutrient broth (1:1,000).

dormant. Fourteen days of storage at 18° produced the same result, and none could be induced to hatch after twenty-one or more days at 18° until a prior conditioning interval was interposed. Further lowering of the temperature below 10° showed little effect on inducing latency, but of course it would continue latency of embryos.

Recovery time for return to an active hatching state is faster for eggs stored at 10° than for those stored at 18° when returned to 25° before hatching (Table 10). When held at 10° for twenty-one days, one-fourth of the eggs could be hatched immediately; over half hatched after seven days, and three-fourths hatched after fourteen days at 25°. Even after fourteen days at 25° only half of the eggs stored at 18° could be induced to hatch.

Ability of embryos to remain latent at intermediate ranges of temperature as well as at low temperatures would have much to do with permitting wide ranges in latitude known for the distribution of this species. A population recovers from the latent state rapidly after exposure to low temperatures to permit it to develop in northern climates, and it is restrained from premature hatching in latitudes of variable but less severe winters. The species even ranges into frost-free zones because a period of cold is not required to condition embryos to hatch.

While unfavorable temperature and a dry substrate may cause embryos to remain quiescent, additional factors must act to restrain hatching in the field at times. Symons *et al.* (1906) noted that eggs hatched erratically and even failed to hatch during the same season that others hatched. Miller (1930) noted erratic hatching in samples bearing eggs that were brought to the laboratory. Filsinger (1941)

TABLE 10

EFFECT OF CONDITIONING AT 25° ON HATCHING OF EMBRYOS OF
Aedes vexans DECONDITIONED FOR 21 DAYS[a]

Deconditioning temperature ± 0.5°	Conditioning at 25° (days)	Eggs involved	
		No. treated	% hatched
10	0	50	24
10	7	50	51
10	14	40	72
18	0	50	0
18	7	50	2
18	14	49	53

[a] All hatching at 25° in nutrient broth (1:1,000).

reported that alternate flooding and drying produced consecutive broods. Only a small percentage (2%) of eggs flooded by cold water during the winter and held until May hatched in experiments conducted by Gjullin *et al.* (1950). Khelevin (1961) noted that hibernant eggs of a palearctic population under water in the early spring did not hatch with the snow-melt species and would not hatch until they were dried and reflooded. The population in central Illinois responds in like manner. A site naturally flooded in early March on one occasion remained under water until June. Samples of soil taken under water in early June showed large numbers of viable eggs *in situ,* and no larvae were found in the area during the interval. Certainly one cause for restrained hatching is that many eggs, especially those deposited early in the summer, become buried to depths that prevent embryos from cracking the enclosing shells (see *Egg: Vertical distribution*).

The question of causes for durability of eggs under field conditions is one that has many answers, because conditions to which eggs may be subjected vary widely. That embryos may survive at least two years is corroborated by laboratory and field observations in varied latitudes, as shown above. The significance of the durability on a field population is open to question where hazards of burying, premature hatching, and predation are present. Few observations have been made where eggs exposed to the vagaries of ecological stresses were recovered at intervals and examined.

Breeland *et al.* (1965) exposed eggs from sources in Alabama and Minnesota and removed them for hatching after exposure of part of the population to winter conditions in each state. Eggs from a Minnesota population showed more viability (75-90%) when brought into the laboratory during the spring after a winter in the field at both sites. The Alabama one showed undiminished survival in Minnesota but decreasing viability outdoors in Alabama.

A field experiment was conducted with a central Illinois population restrained so that it could be recovered at will. Eggs protected from burial and predation survived the vicissitudes of serial winter and summer weather according to source of the eggs. Lots of fifty eggs of strains from Manitoba (55° N.L.), Illinois (40° N.L.), and Alabama (35° N.L.) were used. All eggs came from laboratory cultures three generations removed from the field. Eggs were obtained in late October and were held at 25° for incubation for thirty days. At this time they were separated into lots of fifty eggs each and placed on sifted soil under a thin cover of fragments of oak leaves in glass tubes (dia.: 2.5 cm; length: 7.5 cm). Ends of the tubes were closed with caps of stainless steel screen (40 mesh) to ward off invading arthropods and worms. Viability of all three strains at the time of placement in the field was **94** percent. The tubes bearing eggs were placed on end in a trench in a wooded area well above any zone of flooding. The trench was filled with soil to the level of soil in the tubes. Eggs were removed at intervals after the spring thaw (**150** days), again the following November (**365** days), and finally in late summer (**638** days) after two winters and two summers on the soil.

Results from the exposure are summarized in Table 11. The number of eggs recovered was high (**82%** or better) for the first **515** days of exposure. Survival was also very good but did vary according to the source of the eggs. Eighty percent of the eggs of the Manitoba strain survived two winters and an intervening summer; nearly **70** percent of eggs of the Illinois strain survived; about two-thirds of those from northern Alabama died by the end of the first summer (**337** days). Winter kill was minimal in all strains.

Loss during the second summer (between **515** and **638** days) was caused by two major factors. The tubes were moved to different locations, and the caps on the bottoms were not placed securely when inserted in the soil. Earthworms invaded and disturbed the soil to such an extent that only one to three eggs were recovered from each lot of fifty exposed. Presumably the other eggs were destroyed. In those cages where the soil was undisturbed, many (50% or more in some) of the eggs had hatched in the field, and the empty shells were recovered. Old eggs are prone to hatch with little or no stimulus other than wetting. Presumably the eggs hatched under water while rain was falling during the second summer (see *Egg: Hatching*).

Stresses reducing survival of eggs in natural sites include, in order of significance, burial by siltation or action of burrowing animals (primarily earthworms), decline in viability (after one year), and hatching of old eggs when rain is falling at some time during the summer, even though residual water does not persist.

It is improbable that any eggs at the surface of the soil in natural sites fail to hatch in late spring if they are flooded with water at hatching temperature. Samples of soil bearing eggs brought to the laboratory after the first vernal flood in late May contained **182** intact eggs. Five were induced to hatch; the remainder contained dead embryos. Very likely the viable eggs had been below the surface and were not able to hatch (see effect of burial elsewhere).

Eggs will hatch during extended rainfall under some conditions even while water is running over them. Eggs that are in the second summer when exposed to runoff water are most liable to hatch. Hatching has resulted in the field when eggs were stored above a zone of natural

TABLE 11

SURVIVAL OF EGGS OF THREE POPULATIONS OF *Aedes vexans* EXPOSED IN CYLINDERS OF SOIL SUBJECT TO NATURAL WEATHERING ABOVE THE FLOOD LINE NEAR A WOODLAND LARVAL SITE, CHAMPAIGN COUNTY, ILLINOIS

Exposure to hatching	Days exposed in field	Eggs involved											
		Manitoba population				Illinois population				Alabama population			
		Ex-posed	Re-covered	Survived No.	Survived %	Ex-posed	Re-covered	Survived No.	Survived %	Ex-posed	Re-covered	Survived No.	Survived %
Winter–1	30	98	96	94	96	98	89	88	90	97	89	87	90
Winter–1	120	139	113	99	71	147	139	134	91	197	181	123	63
Fall–2	337	94	89	75	80	93	88	72	78	47	45	16	33
Spring–2	515	48	46	38	80	48	45	33	69	47	46	18	38
Summer–2	638	96	32	30	31	142	28	16	4	141	29	16	11

flooding (see above) and has been induced under laboratory conditions. Samples of debris and soil bearing wild eggs were obtained by pressing a six-inch hollow metal square into the soil, and then digging beneath it to permit removal with minimum disturbance. The surfaces were inclined to about 30° to avoid pooling. They were then pelted with water droplets sprayed from a nozzle in a manner that caused runoff, simulating conditions in the field during heavy summer rains. Within thirty minutes 132 (29%), 46 (12%), and 57 (19%) of the eggs in three samples had hatched. Sixty-nine percent of the remaining eggs hatched when submerged for an additional two hours. This experiment shows that eggs will hatch in runoff water and indicates that such may happen in the field during periods of heavy rainfall in mid-summer, even though they are above pool level at the time. If the slope is sufficient, larvae hatching in such conditions may survive by moving with the water to pool level. No doubt many are stranded and die when the rain ceases.

The season of greatest variability in hatching of eggs in natural sites seems to be early fall. In central Illinois, fall rains may occur in late September or early October before a time of frost. Hatching may or may not be induced. In 1955 a field site under observation was flooded during the first week in October. Larvae were present in abundance (no counts of eggs were made). In 1969 flooding occurred over a site where several hundred eggs per 625 cm^2 were known to occur. None hatched. In fact, no hatching occurred anywhere in Champaign County that year. In 1970 extensive flooding occurred during the last week of September, and eggs hatched everywhere. In this later instance, at the site under special observation all eggs present were known to be of the current season. One sample (225 cm^2) was taken after recession of the water, and ninety-four of ninety-eight eggs had hatched. Possibly the four remaining were below the surface and could not hatch. Climatological data for the area showed that warmer weather prevailed before flooding with warm rains in 1955 and 1970. Two weeks of chilly weather preceded the flooding with cold rain in 1969.

When embryos require one or more preparatory conditions between completion of embryogeny and eclosion, the sequence is called conditioning in this discussion. For eggs deposited outdoors in early summer, minimum conditioning is that of flooding by summer rains and decreasing oxygen level. Summer eggs that are flooded before completion of embryogeny require drying before the minimum series of events. Eggs that have become buried must be exposed by cracking or other disturbance of the soil prior to inundation and lowering of oxygen levels. Eggs that are deposited late in the season may require additional steps in the conditioning process because cold weather has entrained latency.

Mail (1934) found that eggs of a population in Montana deposited during late August and kept continuously submerged failed to hatch when dried and submerged again. When the eggs were placed at $-10°$ for a day or less, many of the eggs hatched readily when submerged later at $22°$. Exposure to temperatures from near zero to $10°$ for five weeks also stimulated eggs to hatch when placed at $27°$ after submergence.

Eggs of a local population when deposited in late summer under natural light in the laboratory required a time lapse to condition them for hatching. Over 1,000 eggs were deposited in the interval August 15-20 by a laboratory colony in the twelfth generation that originated from a single female. The eggs were retained on a moist surface in a moist chamber at room temperature ($24°$), and lots of thirty or more eggs were subjected to hatching stimuli at intervals thereafter until June. Two lots (age one month) that were subjected to hatching procedures yielded 43 percent in one and 68 percent in the other of their larvae a month later, whereas the remainder contained normal embryos that failed to hatch. Two months after deposition, 89 percent of the eggs hatched. In the interval from ten weeks to five months after deposition (through mid-January), 93 to 96 percent of the eggs hatched. Eggs held beyond five months until mid-June hatched more variably (between 76 and 100%). This series of experiments shows that, when eggs are deposited late in the season, a period of low order of hatchability follows for a time when kept at summer ($24°$) temperatures. Within two months at the same temperature hatching returns to that normal for eggs under best conditions.

Eggs of aedine mosquitoes show three orders of hatching response to depression of level of dissolved oxygen. Eggs of some species (Type I) hatch without appreciable depression in oxygen, as do those of *Culex* and *Anopheles* normally. Others (Type II) may be induced to hatch any time oxygen levels are depressed. Those of Type III will not hatch regardless of the manipulation of the oxygen level until after prior conditioning. Summer eggs of *Aedes mariae* (S. & S.) and *atropalpus* and those of *Stegomyia* and *Finlaya* behave in the manner of Type I. Summer eggs of *Aedes vexans* act as Type I for a week or two after embryogenesis; thereafter they respond as do those of Type II. Eggs of *Aedes stimulans* and those of most *Ochlerotatus* behave as

Type III throughout summer and fall conditions and behave as Type II in the spring. Eggs of *Aedes vexans*, when held at temperatures in the 18-23° range, also react as do those of Type III. Those of Type III require conditioning to entrain the hatching response.

Anderson (1968) explains the low order of hatchability that occurs in the multivoltine species *Aedes atropalpus* (Coq.) as being caused by a combination of low (23-27°) rearing temperature late in juvenile life of the females that deposited the eggs, plus reduced exposure to light. Rearing at 23° under a regulated daily exposure to twelve hours of light and twelve hours of darkness inhibited hatching for over 99 percent of the embryos. High temperature (32°) obviated any effect of light. When the day was lengthened to sixteen hours of light, no inhibition of hatching occurred at temperatures between 23° and 32°.

Hatchability of *Aedes vexans* is affected by light acting on the preceding generation also. Wilson and Horsfall (1970) showed that eggs derived from adults exposed at 25° to a daily alternation of twelve hours light and twelve hours dark produced eggs that behaved as Type III eggs when held at 18° for twenty-one days. Adults exposed to light:dark of 18:6 hatched *in toto* at all temperatures.

Certainly new eggs (25 days old) of *A. vexans* that have been subjected to temperatures between 18° and 10° require exposure to a higher temperature before they respond to hatching temperature (Table 10). Eggs embryonated at 25°, then transferred to 10° or 18° for twenty-one days, required two weeks or longer at 25° before half or more of the eggs could be induced to hatch. Few or no eggs could be induced to hatch within seven days when held at 18°, and scarcely half hatched after fourteen days at 25°. Hatching after exposure to 10° was greater than was the case after 18° for zero, seven, then fourteen days at 25°. In neither case would all eggs hatch. A wild population is thereby shown to have latitude in responses to uniform treatment that would have significant survival value.

While new eggs are adversely affected by exposure to cold, such is not the case with eggs embryonated and held for 230 days at 25° (Table 12). From 94 to 100 percent of all eggs hatched, whether the exposure was for one, two, three, or four weeks. Thus eggs of *A. vexans* will hatch normally after at least 230 days from the time of oviposition without regard to temperature between 4° and 30° in the interval. Incidentally, 230 days is the maximum adverse interval most eggs would be subject to under field conditions. In the case of this species, time will eliminate factors that may have indisposed eggs to hatch, and only the terminal sequence of conditioning events (*viz.*, submergence and reduction of oxygen) is required to hatch all exposed eggs.

TABLE 12

EFFECT OF EXPOSURE TO SIX DIFFERENT CONSTANT TEMPERATURES ON HATCHING OF EGG LOTS CONTAINING 50 EMBRYOS OF *Aedes vexans* PREVIOUSLY CONDITIONED BY STORAGE AT 25° FOR 230 DAYS FOLLOWING OVIPOSITION[a]

Holding temperature ± 0.5°	Eggs hatched									
	0 days after storage		7 days after storage		14 days after storage		21 days after storage		28 days after storage	
	No.	%	No.	%	No.	%	No.	%	No.	%
4	0	0	47	94	49	98	50	100	50	100
10	0	0	48	96	45	90	50	100	50	100
18	0	0	50	100	47	94	50	100	48	96
23	0	0	50	100	50	100	50	100	48	96
25	49	98	48	96	50	100	50	100	50	100
30	..	..	48	96	43	86	50	100	50	100

[a] All hatching at 25° in nutrient broth (1:1,000).

Hatching. Hatching of mosquitoes as used here is the act of cracking of the chorion and escape of a larva. In aedine mosquitoes it follows a sequence of conditioning events for eggs of Type II and III. It results from a convulsion of muscles of the confined larva that causes the cranial hatching tooth to rupture the enclosing envelopes. The circular rupture of the chorion occurs about one-fourth the way back from the anterior end (Fig. 16). The shell anterior to the crack becomes a cap that is pushed off. The compressed larva elongates quickly to more than double its length as it literally floats free from the caudal part of the shell. The hatching tooth of a dormant embryo lies in a shallow depression in a mid-dorsal position on the pointed head. This tooth is a sclerous, melanized member in the cuticula at the site of convergence of the two muscles that later are used in vibrating the labral brushes. These muscles contract to deform the head enough to permit the spine to press against the chorion, under tension at the time. The effect is an instantaneous rupture, just as glass tubing breaks in a ring after nicking with a file.

The muscular convulsion that deforms the head of embryos in eggs of Type II takes place in response to a decrease in the level of dissolved oxygen (Borg and Horsfall, 1953). The level of oxygen that initiates the hatching response varies widely according to age and prior history of the eggs and the temperature at the time of hatching. History of the explanation of hatching of aedine eggs based on oxygen deficiency goes back to the work of Hearle (1926). Most of the early observers reported consistent high rates of hatching when eggs had passed the winter in the field or had been kept about a year in sods brought into the laboratory in the autumn and later flooded. Sods brought in at other times of the year yielded larvae erratically. Hearle made the significant observation that flooding sods at 38° accelerated hatching greatly. Rösch (1933) found that eggs would hatch in an atmosphere of ether, but he did not relate the fact to a reduction in oxygen. Gjullin *et al.* (1939, 1941) were the first to publish the principle that a reduction in oxygen could cause aedine eggs to hatch under some conditions. Their deductions were an outgrowth of work on flooding with nonsterilized infusions of leaves and herbal detritus. Unsterilized animal and vegetable products (milk, liver, and amino acids) in water caused hatching. Reducing agents also stimulated hatching. When oxygen was bubbled through the medium, eggs failed to hatch even in the presence of reducing agents such as cysteine.

Work at this laboratory began with the use of nonsterile infusions from plant material, and a consistently high but seldom complete hatching resulted when eggs from natural sites were used. Diluted broth from canned corn was first used as the hatching medium of plant origin. Solutions of certain sugars also stimulated hatching. Because of ease in handling and consistency of results, solutions of a commercial nutrient broth became the routine hatching medium of choice in this laboratory (Horsfall, 1956). The simplest of all media, and one that works well when eggs have been conditioned uniformly, is water in an atmosphere of nitrogen. When nitrogen is used, the dissolved oxygen diffuses from the water into the atmosphere above; as it does so, the level progressively decreases in the water (Horsfall et al., 1957, 1958).

The standardized method for using solutions of nutrient broth at 25° is as follows. A dilution of 1:1,000 is made in unsterilized, aerated, deionized water and placed in a shell vial. The eggs are put in a vial small enough to keep them tightly clustered; the vial is filled with hatching medium and immersed in the same dilution in the larger vial. The eggs are not sterilized and have been shown to be uniformly associated with Pseudomonas spp. and other bacteria. Microbial growth is fostered by the nutrient broth while being incubated in a closed chamber until the oxygen has attained a level sufficient to complete hatching. In the normal course of events, larvae will be out in one to four hours at 25°.

When hatching is promoted by reducing the level of dissolved oxygen by diffusion into a nitrogen atmosphere, conditioned eggs are submerged in a small volume (2-3 ml) of shallow water (2-5 mm deep) in a chamber (1 liter capacity), and nitrogen is forced slowly through chamber for five minutes to replace the air. The system is then closed, and hatching takes place in thirty or forty minutes. Nitrogen causes no adverse effects on newly hatched larvae, even after exposure for an hour or more. By varying the depth of the water and/or the area of surface exposed to nitrogen atmosphere, the rate of decreasing oxygen by diffusion from the water may be varied. When the state of conditioning a population varies, a slow draw-down of oxygen will cause all hatchable eggs to hatch.

Any number of chemicals that remove the oxygen may be used, but they are often toxic to larvae or act to lower the level of oxygen too rapidly or too completely to initiate hatching responses in all embryos. In order to provide both hatching incentive and survival capability, biotic deoxygenation (by microbial growth) is the method of choice as a routine procedure in this laboratory.

The effect of temperature at the time of hatching on rate of hatching of Type II eggs in dilute nutrient broth in tubes (20 mm diameter) is shown in Table 13. At 35° all eggs hatched within eight hours in all

TABLE 13

EFFECT OF DILUTION OF NUTRIENT BROTH ON INTERVAL OF HATCHING OF EGGS
OF *Aedes vexans* FROM A POPULATION REARED IN CONSTANT
LIGHT AND HELD AT 25° FOR 45-60 DAYS

Hatching temperature	Dilution	Eggs treated	% eggs hatched			
			0-4 hr	5-6 hr	7-8 hr	24 hr
35°	1:5,000	45	100	..	..	..
	1:9,000	34	65	32	3	..
	1:13,000	39	48	37	2	..
25°	1:5,000	29	0	0	0	100
	1:9,000	20	0	0	0	70
	1:13,000	30	0	0	0	26

dilutions up to 1:13,000. Most hatching occurred within the first four hours in all dilutions, but some straggling occurred in high dilutions. At 25° much more time was required. Twenty-four hours were required to hatch all eggs in dilutions of 1:5,000; only two-thirds were induced to hatch in dilutions of 1:9,000, and only one-fourth responded within twenty-four hours at 1:13,000. Probably the rate of microbial activity in the higher dilutions was so low that diffusion of oxygen into the medium from the air replaced much of that lost to the microbes in the tubes used. The lower range of temperature at which hatching of post-winter eggs occurs is 14-16°.

Crowding of eggs in the hatching chamber affects both the extent and speed of hatching. Eggs used were held at 24° throughout incubation for conditioning. They were placed in round-bottom vials (5 mm I.D.), which were placed in larger shell vials and then flooded with nutrient broth 1:1,000 for hatching. All (100%) hatched in three to four hours at every range of temperature from 16° to 24°. Eggs in shell vials (15 mm I.D.) only with similar dilutions of nutrient broth hatched as follows: at 24°, 90 percent hatched in five hours; at 20°, 50 percent hatched in six hours, and twenty-four hours were required to hatch 90 percent. At 16° only 30 percent hatched in six hours, and no more hatched within twenty-four hours. In the small, round-bottom vials, eggs were packed together. In the flat-bottom, larger vials, eggs were not touching. Presumably oxygen consumption of the embryos themselves hastens removal of oxygen in the more confined space of the smaller vessels.

Resistance. Hazards commonly affecting eggs of *Aedes vexans* are imposed by drought in summer, low temperature in winter, premature submergence in winter and spring, burial in all sites by burrowing animals or siltation, and predation. Occasional hazards include fires in the overburden of detritus, trampling and burial by herds of herbivores, dislodgement by splashing rain drops, rushing of waters during floods, and premature hatching in runoff water. Size and shape of eggs help prevent injury caused by pressure and by buffeting from current or rain. They readily come to lie in protective crevices or simply roll to one side as hoofs press against them. The nature of the chorion in combination with the size and shape of the eggs resists external pressures usually encountered. No resistance to burial is known, but buried embryos retain hatching capability for a time that may allow many to hatch whenever cracking or curling of the soil exposes eggs to proper hatching stimuli.

Resistance to prolonged cold at or near 0° is routine with little or no loss in hatching ability for a year or more in the laboratory. Eggs have survived up to two years at 4° in the laboratory. Snow cover provides protection from extreme cold in northern latitudes of the range. In middle latitudes, such as that of central Illinois, soil temperature may reach −25° and some injury results. Such temperatures are common at the soil surface for short durations because of the erratic occurrence of snow cover. In one instance during a winter of minimum snow cover, 40 percent of 943 eggs that lay on soil above the water level over winter were not viable in the spring. Mail (1934) subjected eggs to subzero temperature for 160 days without killing embryos.

Resistance to drought is particularly noteworthy. Mail (1934) noted that eggs of a population in Montana, when placed in dry glass vials, withstood 0 to −10° for twenty months and remained viable. Brust and Costello (1969) noted that eggs of a Manitoba population that had been treated to winter conditions for five months and then exposed to an atmosphere of 20 percent RH for fourteen days failed to hatch. About 50 percent hatched when the RH was 50 percent, and 88-98 percent hatched when it was 100 percent RH. Similar results were obtained when a population from Illinois was left on dry cellucotton from March, 1956, to November, 1957, at 8°. Throughout the twenty months the eggs were tested for hatchability, and from 70 to 86 percent hatched. A layer is located between the embryo and the chorion that inhibits passage of water, resulting in resistance to desiccation in dry environments and resistance to uptake of water when submerged.

In drought years survival of embryos may be severely tried. Such was the case in central Illinois in 1953-54. Eggs were deposited by early July, 1953, in a woodland site, after which the site was not flooded again that year. The soil dried quickly after the short periods of rain in the interval July, 1953, to August, 1954. Only 17 percent of eggs taken a year later (July 22, 1954) were still viable.

Embryos survive in eggs after burning of sods (Miller, 1930), because presumably they are actually heated very little. They do not survive well on cattail marshes where the eggs are on muck and the surface is charred, as has been noted in northern Illinois during drought years. They would suffer a similar fate if torches were used to burn detritus.

Water is lost from eggs during storage (Brust and Costello, 1969; Costello and Brust, 1969). A strain from Winnipeg, Manitoba, after twenty-eight days of storage lost over 50 percent of the moisture during a one-day exposure to 50 percent RH. They gained a third of the weight lost in three days at 100 percent RH. In nature eggs are found on surfaces having some moisture and in an atmosphere at or near saturation, so they resist water loss very well.

Embryos withstand submergence for extended periods under laboratory conditions. Eggs have been kept in aerated water at 25° for two months without injury. Another lot was kept submerged and viable in water at 4° for twelve months. Forty-one of forty-nine eggs hatched after seven days at 24° on a moist surface.

Eggs may be induced to hatch in the presence of atmospheres containing partial pressures of carbon dioxide. When eggs were placed in atmospheres containing 10 percent carbon dioxide and 90 percent air, and 50 percent carbon dioxide and 50 percent air, for twelve hours, 37 and 50 percent hatched. There was an immediate inhibitory effect of 100 percent carbon dioxide on hatching of eggs of all ages when exposed for one hour (Table 14). A series of eggs was held at 25° and hatched at weekly intervals after exposure to either 10 or 100 percent carbon dioxide for one hour. They were then held an hour in air and were then subjected to an atmosphere of nitrogen for hatching. An inhibitory effect on hatchability was noted for the first twenty-eight weeks, but after untreated embryos had begun to decrease in hatchability (beyond twenty-nine weeks), no inhibitory effect attributable to carbon dioxide was indicated. Neither was any inhibitory effect attributable to a level of 10 percent carbon dioxide observed.

Mortality factors. Eggs are subject to two significant decimators and several minor ones, but none acts to regulate the species beyond limited focal situations. On flood plains eggs may be buried by silt during winter floods or by other means less understood at other times during the year. During periods of drought extending for two or more years when oviposition sites are not flooded, eggs will fail to hatch.

TABLE 14

EFFECT OF EXPOSURE OF EGGS OF *Aedes vexans* OF DIFFERENT AGES FOR ONE HOUR
TO CARBON DIOXIDE ATMOSPHERES ON HATCHING IN AN ATMOSPHERE
OF NITROGEN ONE HOUR LATER

Age in weeks at 25°	Eggs involved								
	0% CO_2 atmosphere			10% CO_2 atmosphere			100% CO_2 atmosphere		
	Eggs used	Eggs hatched No.	%	Eggs used	Eggs hatched No.	%	Eggs used	Eggs hatched No.	%
3–14	216	202	93	263	247	94	234	159	68
15–28	330	312	95	330	308	93	331	220	67
29–42	362	234	65	357	254	71	339	208	62
43–46	102	19	18	104	16	15	104	12	11
47–52	75	0	0	75	4	5	75	0	0

They fail to hatch under field conditions where flooding occurs during the winter and remain flooded thereafter. Besides burial and lack of submergence as noted above, eggs may be subject to attack by predacious mites and possibly other arthropods, and ingestion by earthworms and incorporation in their dejecta. James (1961) noted that ants (*Lasius* and *Myrmica*) as well as carabids ate eggs of *Aedes* mosquitoes in Ontario.

Larva

Appearance (Fig. 17). Larvae in instar 4 are variable in size and are in the intermediate range of sizes for aedines. Those in vernal pools that develop at about 20° are of maximum size for the species. Those in sunlit pools in mid-summer may mature yet be nearly half the size of vernal ones. Larvae in sunlit pools tend to be light in color; those in woodland ones tend to be dark. The following combination of characters permits recognition of the larvae in Holarctica. Siphonal hair tuft is no longer than width of the siphon at the base and is inserted beyond the pecten. Pecten with one or more apical teeth are set apart from the remainder. The anal segment is not ringed by the saddle. Comb is composed of less than fifteen scales set in an irregular row. Antennae are shorter than the head is long. Head hairs are not aligned with the pre-antennal hair, and the upper head hairs are multiple (2-7) while lower ones are single or multiple (1-6 hairs). Good taxonomic characters are given in Ross (1947), Barr (1954, 1958), Dodge (1966), Bohart (1954), and Nielsen and Rees (1961). Two dark spots appear in the hemocoel of abdominal segment 6 late in instar 4 of larvae bearing the male gonads. Actually the fat body surrounding each gonad is the part that darkens.

Price (1960) has provided descriptions and a diagnostic key to larvae in instar 1 for Minnesota species. In the population of *A. vexans* the one- or two-branched siphonal hair is at the end or barely beyond the pecten. The distal pecten tooth has one long and several short spines. Lateral hair of the anal segment is simple. Antennae are spiculate and have three long apical spines. The postclypeal hair is slightly anterior to the lower head hair. Dodge (1966) concurs.

Sweet and McHale (1970) have cultured cells from macerated larvae in hemolymph-free media. The cells grow in suspension and exhibit a variety of polyploidal cell types.

Focal distribution. Larvae may be present in newly flooded depressions in the ground at elevations under 1,400 m (Baker, 1961) wherever rains provide sufficient runoff to flood them with water under conditions that cause eggs to hatch (see *Focal distribution*). They appear in pools on flood plains as well as those in upland woods, wet prairies, ditches, and canals. In California larvae occur in grassy irrigation ditches without algal mats (Seaman, 1945). They occur in ground pools above high tide along the east coast of USA-48 (Smith, 1904b; Symons *et al.*, 1906). Grassy pools in pastures are prolific larval sites on the plains (Tate and Gates, 1944). In the northern tier of states and southern Canada, shallow, transient pools in meadows, margins of swamps, flood plains, and sandy wastelands provide most larval sites (Twinn, 1931; Owen, 1937; Hearle, 1926). Price (1963) noted that *A. vexans* was the dominant species in woodland pools and was fourth behind *A. excrucians, barri,* and *cinereus* in cattail-sedge marshes. Seepage pools in irrigated areas and river valleys in flood produce this species in desert areas such as parts of Utah (Nielsen and Rees, 1961). Willow thickets provide larval sites in Utah (Rees, 1943) but rarely do so in the Pacific Northwest (Hearle, 1926). In France wet prairies produce larvae in April, and woodland pools do so in summer (Callot and Van-Ty, 1944). The species is a product of river valleys in Czechoslovakia (Trpis, 1962) and wet meadows in Denmark (Wesenberg-Lund, 1921).

Within a specific pool or larger unit such as a flood plain or lake, larval concentration varies according to age of larvae, depth of water, and movement of the water. Distribution tends to be peripheral in all sites for young larvae at least. Such is especially the case on flood plains and in ponds and lakes. When rivers leave their channels in the spring and summer, waters encroach on horizons bearing eggs; the eggs hatch within a short time (usually while the water continues to rise in summer). Young larvae move toward the margin with the rising water, along with floating detritus. Similarly, larvae hatching in rising impounded waters move with the expanding margins of the flooded zone. If the rise of water is rapid and larval drift is unimpeded, larvae may be concentrated within a meter or two of the margin throughout instar 1 or longer. Should the fall in water level be rapid, larvae disperse by drifting. Should the level remain constant or fall imperceptibly, older larvae will disperse actively and often widely. Larvae in instar 4 usually occupy the whole flooded area. As a result of extensive spread late in larval life, little can be learned about their actual origin in large bodies of water. Incidentally regulatory activities based on toxicants are best directed against larvae during the concentrated phase.

The association between margin and distribution of recently hatched larvae is affected by emergent vegetation in the site. A brood appearing in mid-July in central Illinois and originating in a woodland slough some 15 m wide and 60 m long was not strictly marginal in position. While larvae were in the second instar, samples were taken by dipping in a band of vegetation (lizard's tail), open water, and margin free of vegetation. Incidence was 95, 128, and 552 larvae respectively per ten dips taken in these areas. When dense, emergent, herbal vegetation is present, larvae tend to remain near hatching sites and seldom drift to the margins. Such is the case in grassy swales, ditches, irrigated crop land, and flood plains overgrown by dense stands of grass. In one such site larvae were present at a rate of 77/900 cm² over an area of 6,000+ m².

Seasonal occurrence (see also *Adult: Seasonal occurrence*). Larvae may occur in recently flooded sites when water temperature is above about 15°. In sites flooded by water at 10° or lower, no larvae appear (see *Hatching*). Suitable thermal conditions may be met as early as March in central Illinois and any time thereafter until early autumn. During 1956, new larvae were seen each month from May to October in one site. The record of earliest occurrence of larvae in the twenty-year span of these records was March 16 in a site flooded March 14 when the water temperature in the shade was 17°. The larvae subsequently disappeared when the temperature remained 9-12° until after mid-April.

During the interval 1950-70 larvae were seen to hatch in early October twice (1955 and 1970). A record was kept of the brood hatching in 1955. During October 5 and 6 some 130 mm of rain fell, flooding a dry woodland slough near Urbana, Illinois. Larval records were kept from October 7 to 24. Large numbers of larvae in instar 2 were found in margins of zones first flooded. Some larvae were in instar 3 by October 10 and some in instar 4 by October 14. A few pupae had appeared by October 24, even though the water temperature was 12° to 15° after October 17. No adults emerged from this brood. The brood that originated early in October, 1970, emerged as adults but failed to oviposit significantly.

Larval abundance in Minnesota has been shown by Price (1963) to vary widely from season to season. Whereas the usually dominant species (*A. excrucians* Walker) varied by a factor of five from season to season, *A. vexans* varied by a factor of 350. Larvae may be found any month in the year in southern Mississippi (30° N.L.), according to Harden and Poolson (1969).

Kato and Toriumi (1956) reported that the form occurring in Japan appears in ground pools immediately after rainfall when the benthon is dominated by *Difflugia* and the plankton is largely *Cyclops* and *Chydorus*. By the time larvae mature, *Chlamydomonas* dominates the benthon.

Associated species (see also *Egg: Associated species*). *Aedes vexans* may be found as larvae with any of the species coincident as eggs. It also occurs with most species that oviposit on water surfaces when *A. vexans* eggs are marginal to lakes and ponds (Dorsey, 1944). In an area east of Urbana, Illinois, the population of eggs during the winter was composed of *A. stimulans, grossbecki, trivittatus, vexans, sticticus, canadensis, dupreei,* and *Psorophora varipes*. First flooding was partial in February on one occasion, and some *A. stimulans* hatched. Full flooding came during the first week of April when *A. stimulans, vexans, canadensis, grossbecki* D. & K., and *sticticus* hatched. The following year the area was flooded April 28; *A. vexans* was dominant, with *trivittatus, sticticus,* and *stimulans* being rare. Elsewhere in central Illinois *A. vexans* has been collected with *Culiseta inornata* (Will.), *Culex territans* Walker, *restuans* Theo., *pipiens* Linn., *Anopheles punctipennis* (Say), *Psorophora howardii, ciliata, ferox,* and *horrida.* On the plains of west-central USA-48 it occurs with *A. dorsalis, nigromaculis* (Ludlow), *Culex pipiens,* and *Psorophora signipennis* (Coq.) (Tate and Gates, 1944). On the northern plains it occurs with *A. aurifer* (Coq.), *campestris* D. & K., *dorsalis, dianteus* H.D. & K., *excrucians, fitchii* (F. & Y.), *flavescens* (Müller), *intrudens* Dyar, *nigromaculis, punctor,* and *communis* (Owen, 1937). The form in Guam is associated with *Culex* and *Anopheles* species (Reeves and Rudnik, 1951). The form in China (Hunan) occurs with *Culex* and *Anopheles* species (Chang, 1939). In the Danube Valley it occurs with the aedine species present together with *Culex pipiens* and *Anopheles maculipennis* Meigen (Trpis, 1962). In summary, the species may be found with any species which it overlaps in range because of the diversity of sites suitable for oviposition and because of its range in seasonal hatching ability.

Dispersal. Dispersal of larvae occurs in two stages. The first or primary dispersal is away from the hatching site and occurs when larvae in the first instar appear in encroaching water and move passively with the spreading water. Larvae hatch from eggs at lower horizons and mix with those at higher ones as the water levels rise during flood. In small pools primary dispersal is minimal, while on flood plains it may extend over hundreds of meters. Secondary dispersal begins as soon as water stops rising and is obvious when the water begins to recede. In small pools the whole flooded area may be occupied. In valleys of rivers, larvae come to be dispersed through residual pools as the stream returns to its channel, or they may go into the channel and be propelled downstream many kilometers.

Dispersal in the channel has been observed often in Illinois, where records were kept of movement during recession of a flood of the Sangamon River in June, 1960. Larvae were late in instar 4 by the time the river began receding rapidly (about 50 cm a day). The area under observation was a plot of dense canary grass bounded by a broad band of open water where water was flowing toward the channel of the river 200 m away. Larvae were abundant in the channel and moving passively in the current. Larvae in the grass were actively wriggling in all directions, but the drift was toward the more rapidly flowing water. Pupae tended to become stationary on the upstream sides of flotage or emergent vegetation. Once larvae were in the main channel, they tended to become marginal in slack water along any obstruction, in cavities along the margin, and behind exposed roots and near sand bars. However, many seemed to ride the current without obvious resistance.

Feeding. Larvae are omnivores that feed on suspended particulate matter in the first instar, while older larvae browse on matter attached to or lying on submerged objects. Mouth brushes of the young larvae are those of filter feeders, while those of older ones are intermediate between strict browsers and filter feeders. Rudolfs and Lackey (1929) reported that they feed on dinoflagellates, *Chlamydomonas*, and *Euglena* in the laboratory. Live yeast fed daily as deposits on the bottom of rearing pans provides adequate sustenance. When the yeast is dead and serves as substrate for scum-forming microbes, the mass may be destructive of larvae.

Development. Aedes vexans is able to survive and grow during instars 1-3 at temperatures of 11° and above in the field. At 10° larval development required forty-six days, but few survived for a Canadian population (Trpis and Shemanchuk, 1970). Adults emerged from a site in central Illinois by April 28 after hatching April 4. During that time temperatures of 11° to 13° prevailed during instars 1 and 2, 13° to 15° prevailed during instar 3, and afternoon temperatures of 18° or above occurred after April 20. At 20° larvae completed development in ten days (Trpis and Shemanchuk, 1970) when fed live yeast. Brust (1968) found that larvae reared in the dark at 26° pupated within 124 hours when fed daily a diet composed of dog biscuit, live yeast, and blood fibrin (population level: 40 larvae/100 ml of water). Gunstream and Chew (1964) noted that instars 1, 2, and 3 required a mean of sixteen to eighteen hours each and instar 4 some thirty-seven hours at 27.8°. Möhrig (1969) reported that larvae may complete development at 16°; optimum thermal range is 28-30°, and lethal temperature is 35°. Mail (1934) reported that mature larvae could withstand 37° but early instars could not do so. Larvae in the laboratory will survive through instars 1 and 2 at 35°, but few or none pupate at this temperature. Nayar and Sauerman (1970) reared larvae of one aedine (population levels: 75 to 200/350 ml sea water diluted 1:10) to pupation in ninety-eight hours on a dietary regimen as follows: day 1: 160 mgm live yeast; day 2: 20 mgm liver powder; day 2, 3, 4: 80 mgm yeast; and day 5-6: 40 mgm yeast in light:dark cycle of 12:12.

Development of laboratory cultures from populations in North America ranging in origin from Alabama, Illinois, and Manitoba was observed in this laboratory. All cultures were reared at 18° for comparison as to mortality and developmental time. Cultures of the local population were observed additionally at temperatures of 25° and 30°. All cultures were reared in either horizontal pans (water depth: 5 or 20 mm) or in slanted ones where water depths in each pan ranged from 0 to 80 mm. Temperatures were constant (within 1°) and light was constant. All were fed a slurry of live yeast daily and in amounts that avoided scum formation. Populations were 30-40 larvae per pan. Three replicates were used. Exuviae and dead larvae were removed daily, and the progress of development was noted as changes in instar and stage.

Bionomic charts useful for graphically expressing vitality-mortality relationships have been developed in this laboratory. Those for development at 18° are shown in Figs. 18-26. Survival by instars is indicated by height of bars; time of completion of each instar is indicated by width of bars; mortality by instars is indicated by the shaded space above the bars. Degree of synchrony in development is indicated by the amount of lateral spread (tailing off) of the bars at the base caused by retardation of some larvae. The Manitoba strain survived all treatments, matured faster, and was more synchronous in development than the more southern strains at 18°. No appreciable differences may be seen in these qualities between the Illinois and the Alabama strains. Cultures in water 20 mm deep prospered more than did those in either shallow or variably deep water, but this difference was not noticeable for the Manitoba strain. For bionomic charts of another Canadian population at 10, 15, 20, 25, and 30°, see Trpis and Shemanchuk (1970).

Bionomic charts for developmental stages (larval and pupal) of the Illinois strain at 25° are shown in Figs. 27, 28, 29. This strain develops well at this temperature regardless of the water depth. Mortality was virtually nil. The juvenile span was lengthened in the last larval instar in cultures where the water was variably deep but survival was not affected. Pupation began on day 5 of all cultures and was essentially completed on day 6 except for a few stragglers in the shallow water. Emergence began on day 7 in flat pans and was delayed a day in slant pans.

All development was further compressed in time at 30° (Figs. 30, 31, 32). Pupation began during day 4 and was completed during day 5. Emergence began during day 5 and was largely completed on day 6. Mortality was noticeably increased in both deep and shallow water where nutritional stresses seem to be greater. The culture in the slant pans had a mortality rate of 40 percent. Even though development is hastened at 30°, this temperature is not optimal for the species (see also Trpis and Shemanchuk, 1970).

The bionomic charts serve to show at what step in development and to what degree rate of development and mortality act to express degree of desirability of the environment. Considering the charts for the Illinois strain (Figs. 19, 22, 25, and 27-32) on the basis of uniformity of development and minimum mortality at 18, 25, and 30°, the most favorable conditions for laboratory rearing were met in water 20 mm deep when the temperature was 25°. The least favorable ones were in water 5 mm deep and at a temperature of 18°. A temperature of 18° lengthens larval life so much that, in a feral state, larvae would be subject to additional decimation from antagonists such as vorticellids.

Respiration. Aedes vexans obtains dissolved oxygen for respiration by diffusion through the cuticula and aerial oxygen through spiracles in the respiratory siphon. Larvae in early instars readily use dissolved oxygen, and they may survive even when forced to derive their oxygen solely by diffusion through the cuticula. A demonstration of this capability was made in a glass standpipe 300 cm deep and 9 cm in diameter. Larvae were induced to hatch in tubes at the bottom of a standpipe containing aerated water. Young larvae were active enough to reach the surface even after several hours. Larvae in instar 4 were unable to do so at a temperature of 25°. They wriggled up about 100 cm, where they ceased wriggling and settled to the bottom. After several minutes the oxygen debt was restored and the struggle upward began again. Presumably diffusion alone was inadequate to supply the muscles in large larvae during the activity required to attain the surface. Dependence on aerial oxygen has long been the weakness exploited to effect population decimation by covering the water surface with a toxicant.

Excretion. Wastes that pass through the anus are derived from the conspicuous mass of malpighian tubules and the gut. Frequent evacuations occur as semi-solid pellets that fall to the bottom. At 25° pellets may be voided at intervals of fifteen minutes. Ingested material may pass through the alimentary tract in fifteen or twenty minutes in an actively feeding larva. By providing particulate carmine for ingestion by actively feeding larvae for a minute or two, the carmine was ingested and was visible through the cuticula. Its progress could be watched regularly.

Secretion. Salivary secretions are profuse. A mucilaginous exudate is produced by salivary glands that cause food to adhere in the mouth when larvae are feeding on suspended particles. Sometimes saliva is used to trap particles not suitable for ingestion. These adhere in the spines of the galea and are discarded. Larvae no doubt have the usual endocrine secretions associated with development.

Aggregation. Larvae, especially in the last instar, form dense masses of hundreds of individuals under some conditions. Aggregations of older larvae may contain younger ones, but seldom do larvae earlier than instar 3 form massive units. An aggregation is nearly spherical when water is deep enough or is a flattened mass in water less than 10-15 cm deep. Larvae in such masses or balls are in frenzied activity, with each seemingly oriented toward the center of the mass. Aggregations usually appear at places in water which are relatively free of emergent vegetation. Balls disrupted for any reason will reform shortly in the same area or nearby. Not all populations form these balls; indeed, not all larvae in a site at the time join the mass. Possibly the tendency to form balls or to refrain is under genetic control. Nayar and Sauerman (1968) noted similar aggregations of larvae of *Aedes taeniorhynchus.*

Antagonists. Larvae may be attacked by internal and external parasites. Among the internal ones are mermithids (Stabler, 1952 in Pennsylvania; Hearle, 1926 in British Columbia). Laird (1956) reported *Agamomermis* in *A. v. nocturnus* in Pacific Oceania. Trypanosomatids, microsporidians (*Plistophora*), a fungus (*Coelomomyces*), and mosquito iridescent virus have been reported occurring in larvae (Chapman *et al.*, 1966). An internal ciliate was found in larvae in Louisiana (Chapman *et al.*, 1967). Kuznetzov and Mikheeva (1970) found sporangia of *Coelomomyces* in the population in the Russian Far East. An external growth of vorticellids has been noted locally, and the felted mats produced at times obviously impaired larval activity. Morris and De Foliart (1971) reported the total decimation of a local wild population by vorticellids in Wisconsin.

Predators of many sorts feed on larvae. Whether they act in any significant way to depopulate an area is questionable. Minnows and small fish consume larvae, but they are seldom abundant in floodwater. Predacious culicids (Chaoborinae) feed on larvae in Canada (Twinn, 1931, 1931a). Culicinae (*Psorophora ciliata* and *howardii*) may live exclusively on a coincident population of *A. vexans*, but they never seriously depopulate a site, in central Illinois at least. Other casual predators include salamanders, fish, dytiscids, and aquatic Hemiptera.

Resistance. Within any focal population some larvae may be resistant to low temperatures and others to higher ones. Thus, when a site is flooded by cold water in the spring, part of the population survives. Others survive better under summer conditions. Their mobility enables them to move to cooler or warmer portions of a pool when sunlight or shade provides differentials in a focus. Tolerance for partial salinity has been noted by Möhrig (1969). Levels below 8,000 ppm chloride may be tolerated fairly well (Kardatzke and Liem, 1972). At 8,000 ppm mortality of 20 percent, together with extension of larval life, occurred. Levels of 12,000 ppm were lethal.

No resistance significant to toxicants or other regulatory practices has been reported. This species lacks resistance to scum-forming microbes or to those forming flocculent deposits such as appear in sewage effluents. Inorganic solids in suspension, such as silt, are tolerated.

Larvae are tolerant to radioactive phosphorus as H_3PO_4 when exposed to concentrations of 10 μc/l when reared at 24° (Quaraishi *et al.*,

1966). The mean counts per minute per 0.1 mgm dry weight ranged up to 7,864.

Mortality factors. The most potent factors predisposing larvae to decimation are low temperature and stranding caused by runoff and drying caused by lowering of the water table. Larvae that hatch early or late in the year may be retarded in development and die. Very often water rises high enough to hatch many eggs, then falls too rapidly to permit complete development. Shallow woodland pools and flooded river bottoms are particularly subject to such adverse changes. A woodland area near Urbana, Illinois, has a high water table each spring, and larvae hatch in hundreds of shallow pools. None of these larvae matures except in rare years of persistent rain. Biotic agents (see *Larvae: Antagonists*) may cause partial but seldom significant decimation.

Efficacious toxicants of sufficient lethality and variety are available. Toxicants dispensed by abatement districts may be significant decimators in proportion to the infested area toxified. Problems of dispersal and timing of toxification adequate to affect a population cause most difficulty in application. Treatment is required for each brood. Permanent reduction of populations has been achieved largely by rapid dewatering, permanent dewatering, and permanent flooding.

Water in which larvae have been reared is toxic to young larvae, at least in the laboratory. Ikeshoji and Mulla (1970) extracted toxic material from 300 ml of water in which over 1,500 larvae of *Culex* sp. had been held for four or five days. While significant as a mortality factor in the laboratory, such a factor is not sufficiently concentrated in natural sites to affect a species like *Aedes vexans*.

Pupa

Little has been published on the pupal stage. An account by Mail (1934) states that no growth occurs at temperatures less than 10°. At temperatures between 15° and 22°, the stage required three to nine days. Development was completed in one to three days at 27-37°, and 40° proved lethal for 50 percent of pupae examined. The local population regularly completed the pupal stage in one day at 27°.

Adult

Appearance (see frontispiece). To the unaided eye this mosquito may appear yellowish because the hairs and scales of the mesonotum are tawny. At other times females are very dark because of black scales on legs and abdomen. Its size is intermediate between the smaller *Aedes cinereus* and the larger *A. stimulans,* two common associates in Nearctica. Minute examination reveals a narrow band of whitish scales on the base of each tarsal segment, particularly evident on those of the hind tarsi of both sexes. Length and distinctness of the bands may vary with population, season, and age of the mosquitoes. Negative, diagnostic features of the female are absence of scales or hairs on the lower part of the mesepimera, and absence of a ring of lighter scales on the proboscis. Wings of females of the oriental variant appear speckled on portions of the subcosta or other main veins because of light and dark scales. The last segment of tarsi of legs 2 and 3 is dark apically. Male terminalia have blade-like dististyles that taper beyond the middle to blunt ends (Fig. 33). Insertion of the dististyles is subapical. Claspettes are fused to the basistyles, are crowned apically with spinose setae, and lack filaments. Relative weights of males and females were determined by Paulovova (1967). For other descriptions see Barr (1954), Gjullin (1937, 1946), and Ross (1947).

Abnormal appearance has been reported in the literature (Cupp and Fowler, 1969). Two males were reported with anomalous features that normally differentiate the sexes. The gonapophyses and tract were those of a male. One had two ovaries; the other had only one gonad (ovary). One had one cercus (♀) and one paraproct (♂) side by side (Fig. 34). One had demasculinized antennae; the other had one male and one female antenna. Whether these mosquitoes were gynandromorphs in the traditional sense or were anomalous males partially feminized by epigenic pressures was not determined. Minson (1969) found a specimen in Utah which was male anteriorly and female internally and externally at the caudal end.

Focal distribution. Aedes vexans is a mobile species that shifts position to occupy several sites serially. Shortly after emergence both males and females may be present in the grass and herbs marginal to larval sites before departure. At this time the males may take to the arboreal canopy during the night. In absence of an arboreal canopy, males may form transient swarms above the herbal layer, especially at dusk. Females during the feeding phase may concentrate for a week or two in low vegetation within and near urban centers, farm buildings, and pastures for livestock. They concentrate in herbal shade provided by foundation plantings, flower beds, and dense vegetation in domestic situations. In feral sites, they congregate in herbal shade in sites free of sylvan canopy and in understory of woodlands. Herbal layers such as that provided by stinging nettle (*Urtica*) and horse mint (*Monarda*) are conducive to concentrations in Illinois.

Post-emergence massing prior to departure consists of accumulations of adults of both sexes in the herbal canopy over or marginal to shallow emergence sites. Twenty to thirty adults per 625 cm² have been noted on a grassy flood plain at such times. They may shift position when a person walks through the occupied area. In summer post-emergence aggregations may persist for one or two days before the dispersal phase. The spring brood may linger in the vicinity of larval sites for several days when persistent low nocturnal temperatures inhibit massive dispersal. Under such conditions they spread throughout the understory and herbal canopy, becoming more numerous in the drier parts of a meadow, woodland, or slopes proximal to the site of origin. A few females may remain near oviposition sites throughout their lives. At a site on the Sangamon River where this was the dominant species as larvae, the post-dispersal biting population near the flood plain was less than 1 percent of the number that emerged.

Gravid mosquitoes en route to oviposition sites across terrain devoid of sylvan canopy in central Illinois may stop in meadows and dense herbal canopy such as that provided by fields of alfalfa. One such situation over 3 km from nearest woods was an aggregation site during late May. Between early June and mid-August none was taken in alfalfa fields.

In Wisconsin in mid-summer adults were annoying and about three times as abundant in areas of low herbal canopy (prairie) as they were in an area of sylvan canopy (Thompson and Dicke, 1965). Peridomestic situations were sites of major accumulation of biting adults in Wisconsin (*idem*). Situations of greatest accumulation in urban Illinois include low herbal canopy of flower beds and clumps or rows of compact shrubbery such as provided by arborvitae, privet, lilac, yew, forsythia, and weigelia. Canopy shade of ragweed, brambles, stinging nettle, and tall grasses provides attractive cover near feeding sites. Instances of adults entering dwellings on dark days have been reported in eastern Canada (Twinn, 1931); however, they do so as transients and do not remain as some *Anopheles* and *Culex* often do. When a population of *A. vexans* is moving, they enter long, low, well-lighted buildings, such as those of assembly plants operating at night.

Seasonal occurrence. Over its range in Nearctica adults of *Aedes vexans* may be present from May through September. In southern USA-48 they may be present earlier and later in the year. Numbers present at any one time will vary according to area and frequency of flooding of larval sites. During seasons of erratic rainfall, adults may be present for only part of the season. Climatological conditions that restrict populations are either seasons of normal or smaller amounts of rainfall without deluges or seasons of prolonged drought with no deluge. In central Illinois a general late-summer drought is normal and of such long duration that local production of adults usually ceases. Even in drought years a general rise in the adult population usually occurs in September, for reasons discussed in the section on dispersal. Smith and Love (1956) showed that *A. vexans* in Georgia is more abundant in a season of erratic, low rainfall than in one of high rainfall. During the summer of high rainfall 1.2 percent of a light-trap catch was of this species, while over 94 percent of a similar collection was of this species during a dry season. The numbers obtained in traps varied drastically from a low of 55 for the wet season to 7,712 for the dry one. A low incidence is the rule in central Illinois when precipitation of about 15-25 mm falls in an hour at weekly or biweekly intervals during the summer.

Adults are variably abundant in a specific locality according to causes for aggregation. Near a source they are in maximum numbers as both sexes and are concentrated in the herbal layer for a short time after each interval of emergence. Females are again common in such sites, but they are less concentrated as gravid ones invade for oviposition. Incidence rises precipitously or gradually at feeding sites (urban, peridomestic, or sylvan), according to productivity and proximity of juvenile sites, daily rate of departure during dispersal phase, and suitability of shelter provided in aggregation area.

Records of seasonal abundance of mosquitoes based on trapping at light such as are kept by abatement districts provide useful data on local numbers. Light traps usually provide information on intraspecific variations, but their results may be misleading when interspecific comparisons are needed (Huffaker and Bach, 1943). Even quality and intensity of light (up to 75W) may be at variance with actual population densities (Masters, 1962). Herewith are given records of populations of *A. vexans* during the summers in three parts of Cook County, Illinois, for the interval 1960 to 1969 based on light traps. The traps were located in suburban, sylvan situations north, west, and south of Chicago. Results are shown in Figures 35, 36, and 37. While the figures show only numbers in excess of an average of twenty mosquitoes caught per night at each of the three locations, mosquitoes were present in all areas during the summer months. Adults appeared each year in May, and a few were taken as late as October in some years. The lean years were 1960-64 for areas north of the city, 1960-63 for suburbia south of the city, and 1960-66 for areas west of Chicago. From 1966 to 1968 the mean number collected per night increased, and the number of months showing higher incidence increased on the north and south sides. The numbers collected on the west side were uniformly less than the other areas, but trends were similar. During the ten years reported herein, the number of nights when the maximum collections at one trap in a sylvan location in each area exceeded 100 mosquitoes were 41, 6, and 126 for north, west, and south sides respectively. All traps were installed near the periphery of areas where abatement was practiced and are presumed to reflect immigrant populations.

Months of maximum population during any season vary from year to year. July was the month in which aggregate populations were highest most frequently during the decade. July was the month when most nightly collections in excess of 100 females were made. High nightly collections during the summer reflect population densities originating after a period of local rainfall adequate for producing a large brood. This situation prevails when repetitious rainfall causes the water table to rise progressively, local downpours flood different areas to capacity, and deluges flood extensive areas. Repetitious rainfall usually results in few or no large collections in traps on any one night, but such conditions increased the mean monthly collections. Locally heavy rains produced large broods, and trap collections from immigrant sites (as shown in Figs. 35, 36, and 37) were high or lower according to the frequency and extent of the inundation. Local incidence on any one night may be extremely high and probably reflects a moving population. When general flooding occurs in any extensive area, increased incidence occurs in all areas. Such was the case in the interval 1967-69 in Cook County.

Mulhern (1936) reported that abundance of *A. vexans* rose sharply between 1932 and 1936 in New Jersey. Incidence based on light-trap records rose ninefold in Morris County and twelvefold in Salem County. Changes in local flooding caused by local rains or rains in the watersheds above the sites caused local fluctuations. Incidence in three localities in New Jersey reflect these effects. In one locality unseasonably heavy rains in September following drought conditions in July and August produced a sevenfold increase. The same conditions sent the incidence of *A. vexans* in one urban situation up 180-fold.

Incidence of adults moving near the flood plain of the Sangamon River in Piatt County, Illinois, varied widely during the span of three years (Table 15). Stationary interceptor traps (Fig. 38) designed to catch mosquitoes in flight indicated a large population throughout June and July of 1958, few to almost none in 1959, and a moderate abundance in 1960. Precipitation records for the area providing water for

TABLE 15

VARIATIONS IN SEASONAL INCIDENCE OF *Aedes vexans* IN STATIONARY INTERCEPTOR
TRAPS IN PIATT COUNTY DURING THE SUMMER, 1958-60

Interval	Number of mosquitoes trapped								
	1958			1959			1960		
	Total	♂/ night	♀/ night	Total	♂/ night	♀/ night	Total	♂/ night	♀/ night
June 19-23	19	2	1	0	0	0	47	0	9
June 24-28	405	52	29	0	0	0	11	0	2
June 29-July 3	150	11	19	5		1	60	<1	11
July 4-8	72	2	12	0	0	0	16	1	2
July 9-13	15	0	3	1	0	<1	143	1	27
July 14-18	52	<1	13	1	0	<1	7	0	1
July 19-23	136	11	35	13	0	2	8	<1	1
July 24-29	225	7	37	3	0	<1	13	0	2

the broods show over 37 cm of rain for June and July, 1958, 6 cm for
the same interval of 1959, and 30 cm for those months in 1960. The
river failed to flood its valley during 1959 but did flood it two or more
times during both 1958 and 1960.

The numbers of adults trapped at light varied widely over four
seasons in a woodland area that produced large populations near
Urbana, Illinois (Table 16). During 1957 and 1958 a trap (Fig. 39)

TABLE 16

INCIDENCE OF ADULT *Aedes vexans* ATTRACTED TO LIGHT IN A WOODLAND THAT WAS
A SOURCE AS WELL AS AN AGGREGATION SITE ADJACENT TO URBANA, ILLINOIS

Month	Mean number of mosquitoes trapped per night							
	At larval source				130 m from source			
	1957		1958		1959		1960	
	Males	Females	Males	Females	Males	Females	Males	Females
May	13	13	10	11	..	..	1	5
June	176	134	14	12	6	6	3	5
July	51	48	25	28	3	6	4	6
August	..	..	31	29	2	2	..	..
May-Aug. rainfall (cm)	45		55		27		36	

was operated under sylvan canopy surrounded by shrubs. During 1959 and 1960 it was placed in open woods free of shrub layer. Ample rain fell to flood the larval site at least once during each summer. Data in Table 16 indicate that significance of the area as an aggregation site varied from year to year. The place of dense vegetation was more heavily populated than that lacking an understory. In the area of maximum aggregation for the years 1957 and 1958, the numbers trapped varied according to season. Reason for the obviously higher incidence in 1957 was that horses for a riding stable were then pastured in the woodland, while they were absent in 1958. Incidence therefore varies between seasons and between areas within a site because of variations in attractiveness as a site for aggregation.

At Urbana, Illinois, over a three-year period (1956-58), peak periods of abundance were noted each season. In 1956 the first peak occurred in the first week in June; the second, by far the larger, came in early September. In 1957 the peaks came in early June and late June, with a large residual population remaining during July. Four peaks came in 1958 during June, July, and August. Certainly the rainfall during 1958 came in such quantity and at such intervals as to explain the broods. During other years the population rose largely because of invasion from flood plains elsewhere, because local sources were not productive.

Adults of *Aedes vexans* are more active before midnight than later, a fact that is more obvious in September than earlier in the summer. Light traps rigged to separate nightly collections into hourly increments (Horsfall, 1962) were operated during two successive summers proximal to the flood plain of the Sangamon River (Piatt County, Illinois). During the months of June, July, and August 59-69 percent entered the traps between 10 P.M. and midnight (dark hours 1-4), and 31-41 percent entered after midnight. During September 80 percent were trapped before midnight.

Mosquitoes appear as adults outdoors in late April and continue into November in central Illinois. *Culiseta inornata* and *Culex* species appear first as they emerge from hibernation. Additionally they occur latest in the season and are active into November. *Aedes vexans* may mature in April and appear first in traps early in May, because they must come from larvae of the current season. A rotary interceptor trap operating continuously from April until early October during 1956 in a wooded aggregation site collected *Aedes vexans* as early as May 12 and continued doing so until the end of September and sporadically until early November (Table 17). This was a season of subnormal rainfall at the site, except for July, so that the collections produced only a modest population. The flood plain of the Sangamon River (that

TABLE 17

INCIDENCE OF THREE SPECIES OF MOSQUITOES IN AN AGGREGATION SITE BASED ON COLLECTION
BY A ROTARY INTERCEPTOR TRAP NEAR URBANA, ILLINOIS, 1956

Collecting interval		Numbers collected per night												
Date	No. nights	Aedes vexans				Culex pipiens				Aedes trivittatus			
		Male		Female		Male		Female		Male		Female	
		Max	Mean	Max	Mean	Max	Mean	Max	Mean	Max	Mean	Max	Mean
April 21-May 11	19	0	0	0	0	0	0	0	0	0	0	0	0
May 12-22	11	14	4	8	4	0	0	1	0	1	0	0	0
May 23-25	3	13	9	18	4	0	0	0	0	0	0	0	0
May 26-28	3	4	3	15	12	0	0	0	0	0	0	0	0
May 29-June 7	10	8	2	10	4	0	0	0	0	1	0	1	0
June 8-12	5	23	15	13	10	3	1	2	0	5	2	2	1
June 13-July 9	27	7	1	10	4	4	0	1	0	1	0	6	1
July 10-12	3	4	1	43	21	2	2	5	3	0	0	0	0
July 13-Sept. 16	66	8	1	16	5	21	7	14	6	6	0	5	1
Sept. 17-18	2	26	23	18	17	14	14	6	3	0	0	0	0
Sept. 19-20	2	5	3	7	5	8	8	2	1	0	0	0	0
Sept. 21-26	6	38	16	42	23	37	21	11	8	0	0	0	0
Sept. 27-Oct. 8	12	8	3	17	9	41	24	14	7	0	0	2	0
Oct. 9-29	21	2	0	8	2	194	64	49	28	0	0	0	0

approaches the collecting site within 16 km to the west) was inundated during May, June, and early August, producing three major broods. These possibly provided most of the infestation at the collecting site. *Aedes trivittatus* and *Culex pipiens* were of proximal origin. *Aedes vexans* appeared after May 12 and reached major abundance in May, early June, mid-July, and late September. All but the first peak were results of invasion. Possibly the invasion of late September was one from far-away sites (see *Adult: Dispersal*).

Associated species. Aedes vexans has a wide range of resting sites, sources, and times of occurrence over its range. As a result, it may be present with other mosquitoes originating at more restricted times and places. In Illinois it has been found with all of the fifty or more species known to be present in the state. In central Illinois, interceptor traps have collected eighteen species coincident with *Aedes vexans*. Table 18 shows the relative incidence of species present and moving in a sylvan aggregation site that were obtained by a rotary interceptor.

While several species may be associated in any one place in the course of a season, relative incidence changes. *Aedes vexans* was the dominant species in the aggregation site near Urbana, Illinois, throughout most of 1956. *Culex pipiens* rose in abundance so that by September it became dominant and remained so during October in collections obtained by the interceptor trap. Others of the genus *Culex* were abundant, with *C. restuans, salinarius* Coq., and *territans* being the more common. *Culex tarsalis* Coq. appeared first on August 12, but occurs in this locality as a rare species.

Interceptor traps operating at three rural peridomestic locations near the flood plain of the Sangamon River in Piatt County, Illinois, from June 19 to July 30, 1958, showed *A. vexans* dominated the population (Table 19). It occurred with nineteen other species and comprised 73 percent of the collection of males and 54 percent of the females. *Culex pipiens* and *A. trivittatus* together compromised 28 percent of the collection. Two of the prairie species *Psorophora* (*confinnis* and *discolor*) were common. *Anopheles punctipennis* was far more numerous than *A. quadrimaculatus*. *Culex tarsalis* began to appear late in the interval. Stage *et al.* (1937) noted that *A. vexans* dominated the population arising from the flood plain of the Columbia River (northwestern USA-48) in years of low flood crests and that *A. sticticus* did so in years of high crests.

Aedes vexans were attracted to hosts coincidentally with a number of other mosquitoes. In sylvan sites of central Illinois they were collected with *Aedes canadensis* in mid-June between 6:00 and 7:00 P.M. Ratios between the two varied from 1:1 to 1:6 from day to day or

TABLE 18

INCIDENCE OF FEMALE MOSQUITOES[a] CAUGHT IN A ROTARY INTERCEPTOR TRAP AT A SYLVAN
AGGREGATION SITE NEAR URBANA, ILLINOIS, APRIL-NOVEMBER, 1956

Trapping interval	Trapping nights	Aedes vexans	Culex pipiens	Culex salinarius	Culex restuans	Culex territans	Uranotaenia sapphirina	Aedes trivittatus	Anopheles punctipennis	Culex erraticus	Anopheles quadrimaculatus	Orthopodomyia signifera	Culiseta inornata	Culex tarsalis
April 21-May 14	21	10	0	0	14	8	0	0	3	0	0	0	3	0
May 15-June 10	26	158	0	4	51	21	5	5	3	0	0	2	0	0
June 11-July 1	12	65	4	27	34	9	7	17	8	2	0	0	0	0
July 2-Aug. 5	35	216	145	155	23	12	16	38	6	4	4	7	0	0
Aug. 6-Sept. 30	55	433	358	108	117	86	77	35	26	19	10	5	0	2
Oct. 1-Nov. 6	37	103	805	11	35	80	2	0	22	0	1	0	3	3
Total	186	985	1,312	305	274	216	107	95	68	25	15	14	6	5

[a] Others collected rarely and as males only: *Aedes sticticus* and *triseriatus, stimulans, canadensis, Psorophora horrida, confinnis,* and *Anopheles barberi.*

TABLE 19

ASSOCIATED SPECIES CAUGHT IN INTERCEPTOR TRAPS NEAR THE FLOOD PLAIN OF THE
SANGAMON RIVER, PIATT COUNTY, ILLINOIS, JUNE 19-JULY 30, 1958

| Species | Mosquitoes collected | | | | | |
| | Males | | Females | | Total | |
	No.	%	No.	%	No.	%
Aedes vexans	687	73	1,302	54	1,989	60
Culex pipiens	50	5	457	19	507	15
Ae. trivittatus	137	15	308	13	445	13
Psorophora confinnis	38	4	134	6	172	5
Anopheles punctipennis	19	2	46	2	65	2
P. discolor	0	0	41	2	41	1
Ae. sollicitans	0	0	23	1	23	<1
P. horrida	1	<1	22	1	23	<1
C. territans	0	0	18	<1	18	<1
C. tarsalis	0	0	12	<1	12	<1
An. barberi	2	<1	8	<1	10	<1
Ae. sticticus	0	0	8	<1	8	<1
Orthopodomyia signifera	0	0	6	<1	6	<1
P. ciliata	1	<1	4	<1	5	<1
C. restuans	0	0	3	<1	3	<1
Uranotaenia sapphirina	0	0	2	<1	2	<1
P. howardii	0	0	1	<1	1	<1
An. quadrimaculatus	0	0	1	<1	1	<1
P. ferox	0	0	1	<1	1	<1
Ae. triseriatus	0	0	1	<1	1	<1
Total	925	100	2,398	100	3,333	100

site to site. Other species present included *A. stimulans, triseriatus*
(Say), *trivittatus, Psorophora ferox, Anopheles punctipennis,* and
Culex erraticus (D. & K.). In coniferous forests of northern USA-48,
A. vexans may occur in small numbers where *A. communis* and *punctor*
are dominant.

Harden and Poolson (1969) noted that *A. vexans* in southern Mississippi comprised less than 10 percent of the mosquitoes trapped, whether
caught in a New Jersey light trap, CDC light trap, or by a mobile
interceptor (truck). *Anopheles crucians* Wiede. and *Culex salinarius*
were the dominant species with the latter, comprising 61 percent of
the population caught by interceptor and 35 percent of that caught
by CDC trap. *Aedes atlanticus* came to the CDC trap more readily
than to either of the other traps.

Dispersal. This mosquito disperses varying distances after emergence from larval sites. The extent to which a brood may be disseminated is influenced by conditions of light and weather following emergence, direction and speed of air mass bearing them during flight, consistency and direction of orientation of individuals in the air mass, and geographic position of the brood. Conditions at the time of emergence have much to do with height of upward flight before leveling out. The higher the flight pattern, the greater the probable dissemination. The greater the wind velocity, the farther a high-flying population may be drifted. The more coincident the direction of flight and direction of drift of the encompassing air mass, the more concentrated the brood during and after dispersal.

Reports of distances traversed by broods indicate wide-ranging capabilities. Stage *et al.* (1937) found females 24-48 km away after overland flights from any possible larval sites. Möhrig (1969) reported flights of at least 48 km in western Europe. Clarke (1943) marked mosquitoes by staining them while aggregated in emergence sites and traced dispersion of individuals up to 22+ km overland near Chicago, Illinois. One female was collected 10 km from the emergence site two hours after being marked. Headlee (1918) reported similar flights for *A. sollicitans* inland from salt marshes in New Jersey. Bresslau (1917) stated that the spring brood appears about mid-May in Germany but does not disperse widely. Dispersion is by low-level flight resulting in a creeping spread.

During late summer and early fall, when large air masses move across areas occupied by newly emerged mosquitoes, flights of hundreds of kilometers seem to occur (Horsfall, 1954). In one instance a massive invasion into central Illinois accompanied the intrusion of a vast, fast-moving air mass from more northern areas in early September, 1953. A prolonged summer drought over all of the invaded area precluded local origin for the brood. An extensive area in southwestern Wisconsin which produced a massive brood was the closest probable source. The invasion was preceded by unseasonably hot weather during the last weeks of August and early September in Wisconsin, according to Climatological Data 58(9):122. August rainfall in the producing area was 10 cm above normal, while that in Illinois was the second lowest on record. On September 4 a massive intrusion of air from the northwest occurred. Mosquitoes were suddenly annoying along all rivers in Illinois by September 5. A possible movement of 320-800 km is indicated, because the brood appeared in annoying numbers as far south as the junction of the Ohio and Mississippi rivers. A similar flight occurred on September 12, 1959. Again no local source for the flight could be found, yet the number of biting females (without accompanying males) increased from none being collected from human bait to 5 ♀ /3 min exposure (forearm) at one site (wooded flood plain). A nearby light trap operating coincidentally showed an increase from 1 ♀ per night to 40 ♀ at the time of invasion.

Dispersal flights over open water were noted by MacCreary and Stearns (1937). Females were taken in light traps 13+ km from land in Delaware Bay. Curry (1939) reported that *Aedes sollicitans*, a species with similar propensities for dispersal, appeared on board a ship 175 km from the nearest point of land in the Atlantic east of USA-48.

Summertime dispersion is likely to be more circumscribed than is the case in early fall, for example. Air masses in Illinois are likely to be smaller cells in summer and are more likely to dissipate quickly. This seems to be the case where *A. taeniorhynchus* (Wiede.) disperses in coastal Georgia (Bidlingmayer and Schoof, 1957). Ninety percent of a summer population of marked females of this species were retrieved within 6 km of the release point. Males, too, remained near the release point, but recoveries were made up to 19 km distant.

Evidence exists that displacement of populations leaving larval sites is directional. Lighted horizons created by urban communities seem to cause nocturnal flight patterns toward the lights. Clouds and inversion layers of smog reflect light from urban areas which create differential lighting on a horizon visible for miles. Vegetative patterns act as guides for orientation of mosquitoes in flight. Rees (1943) observed flights in semi-arid areas of western USA-48 moving along water courses where vegetation was greener and humidity higher. Klassen and Hocking (1964) reported similar orientation for northern aedines along riverine valley systems in Alberta, Canada, near Edmonton. Multidirectional flight has been observed by Stage *et al.* (1937) in the Pacific Northwest.

Features of the terrain influence orientation of individual mosquitoes during massive flights when there is visual contact with the ground. Movements in the direction of rows of corn have been noted in central Illinois. Flights through alfalfa and soybeans close to the ground have been noted when humidity in air above such vegetation was low. Kennedy (1939) noted upwind orientation of an aedine mosquito being maintained in a laboratory device as long as the objects passed from front to back of the eyes. Little is known about flight at night or at heights too great for visual contact. Differentially lighted horizons cause choice of the lighter one, apparently without regard to wind direction.

Flight of engorged mosquitoes is probably multidirectional (Edman and Bidlingmayer, 1969). Adults that had fed on bovine blood moved

3 km to an isolated oviposition site inside a marshy area in Indian River County, Florida.

Dispersing mosquitoes may fly at various levels from the ground to great heights. Glick (1939) collected females of *A. vexans* at heights of 140, 280, and 1,470 m in air-borne interceptor traps. McCreary (1941) found the ratio of females collected by light trap at heights of 1.3 m and 24 m to be 1:4.

Aedes vexans moves across country certainly up to 15 m above ground. Both males and females were taken in a rotary interceptor trap on top of a six-story building in Urbana, Illinois, between June 10 and July 31, 1957. During this period 9 ♀ and 6 ♂ were trapped. Associated species at this height were *Culex pipiens*, *salinarius*, and *restuans*, *Aedes triseriatus*, *Uranotaenia sapphirina* (O.-S.), and *Psorophora confinnis*. The site of the building was removed from trees and was at least 2 km from any possible source of these species.

Aedes vexans remains close to the ground when flying about in the vicinity of larval sites. Stationary interceptor traps operating at heights of 2.1 and 4.2 m near the Sangamon River for forty-four nights collected 295 ♀ at the lower and 56 ♀ at the higher level, a ratio of nearly 6:1. During five nights when major flights were noted, 180 ♀ were caught at a height of 2.1 m and only 21 at 4.2 m, a ratio of 13:1. Burgess and Haufe (1960) trapped *A. vexans* when attracted to a striped cylinder rotating at 1.5 m over grassland.

At an aggregation site *Aedes vexans* is more abundant in flight at a height of 1.2 m above the ground than it is at 2.4 m, according to collections made with rotary interceptor traps. During 1956 two traps operating at each level during the summer collected 490 ♂ and 985 ♀ (Table 20). Of these 328 ♂ (67%) and 785 ♀ (80%) were taken at the lower level. Not only was the number collected at the lower level higher, but only rarely was the situation reversed. At the same time, in the case of *Culex pipiens* 2,680 ♂ and 1,312 ♀ were collected. Seventy-four percent of the males and 70 percent of the females were taken at the lower level.

The relation between production of a brood and its invasion of an urban center (Urbana, Illinois) was determined during the summer of 1957, a season when rainfall was above average. Rotary interceptor traps were operated nightly in the backyard of a dwelling in Urbana. *Aedes vexans* was collected each night at levels of 1.2 m and 2.4 m between June and early August. The criteria used as evidence of invasion of a new brood was a sudden rise in incidence of males plus the presence of females with scales on thoraces, wings intact, devoid of blood, and not gravid. The first invasion of 1957 began April 29 by low-level encroachment through vegetation. By May 4 adults were present over the city. The June flight of the season arrived from June 27 to June 30. This brood resulted from larvae produced by a general heavy rain on June 13 which caused partial flooding of the Sangamon River Valley to the west and partial flooding of sites nearer Urbana.

TABLE 20

COMPARATIVE INCIDENCE OF *Aedes vexans* AND *Culex pipiens* FLYING AT HEIGHTS ABOVE THE GROUND IN A WOODED AREA NEAR URBANA, ILLINOIS, SUMMER OF 1956

Weekly interval	Number of mosquitoes collected by interceptor trap							
	Aedes vexans				*Culex pipiens*			
	Height: 1.2 m		Height: 2.4 m		Height: 1.2 m		Height: 2.4 m	
	Male	Female	Male	Female	Male	Female	Male	Female
Apr. 24-30	0	0	0	0				
May 1-7	0	0	0	0				
May 8-14	4	8	3	2	1			
May 15-21	19	22	9	7	0			
May 22-28	26	36	11	6	0			
May 29-June 3	6	20	1	8	0			
June 4-10	39	50	16	9	1			
June 11-17	27	23	9	10	5	3	2	1
June 18-24	2	12	1	1	1	0	0	0
June 25-July 1	19	14	5	5	0	0	4	0
July 2-8	6	38	2	5	3	1	2	1
July 9-15	3	81	5	12	8	21	5	5
July 16-22	6	34	3	7	19	28	9	12
July 23-29	2	18	5	6	14	29	7	9
July 30-Aug. 5	6	13	4	2	11	30	6	9
Aug. 6-12	6	21	3	2	20	36	9	14
Aug. 13-19	8	16	3	6	36	40	11	16
Aug. 20-26	7	25	1	4	65	21	22	13
Aug. 27-Sept. 2	3	25	0	9	50	39	11	14
Sept. 3-9	9	42	5	13	39	34	24	11
Sept. 10-16	12	44	6	11	83	32	36	13
Sept. 17-23	72	101	39	29	84	22	34	10
Sept. 24-30	33	64	19	21	87	26	34	17
Oct. 1-7	13	47	9	19	137	38	52	25
Oct. 8-14		16	3	3	126	39	49	19
Oct. 15-21		7	0	0	276	134	111	54
Oct. 22-28		5		2	606	230	149	103
Oct. 29-Nov. 4		3		1	263	96	104	40
Nov. 5-8		0		0	34	12	30	15
Total	328	785	162	200	1,969	911	711	401

The third invasion came on July 22 and involved a brood produced in the environs of Urbana after all known local sites were wholly in- undated earlier in the month. A coincident upsurge of males and young females was noted in the collection made during the night of July 22-23. A heavy rain fell in the interval between midnight and 2:30 A.M. Collections made prior to 10 P.M. yielded no new females. Therefore the arrival time was sometime after 10 P.M. and before 6:00 A.M. July 23. Observations of shrubbery and other vegetation before 8:00 A.M. July 23 showed marked increase in activity of mosquitoes. Adults were flying against the walls of the house in an agitated fashion as if still in transit. Later in the day adults were abundant in the shrubbery but were no longer in an agitated state. Cross-town move- ments may take place within a day's time (Rees, 1943).

The two examples of mass movement indicate the following: a sum- mer brood may be expected to invade urban centers nine to twelve days after larval sites have been flooded; exodus may be stimulated when rain coincides with the urge to disperse; degree of concentration at arrival sites is related to distance and direction from larval sites as well as to direction and velocity of wind during dispersal. An early spring brood or a fall brood may fail to disperse or may do so less dramatically because movement is less concentrated in time. Certainly interceptor traps reflect the time of movement more reliably than do traps based on any principle of attraction (light and bait). Provost (1960) reported similar results with *A. taeniorhynchus* arising from Florida salt marshes. Clarke and Wray (1967) noted that arrival time in urban collecting sites began on day 4 and peaked on day 5 after emergence.

Movement of female *Aedes vexans* is stimulated during nights of major rainfall (Table 21). Stationary interceptor traps operating nightly at one site near a flood plain of the Sangamon River indicated five peaks of activity. Over 25 mm of rain fell during each of these nights. A total of 182 females (36/night) were caught during nights of heavy rainfall; few or no mosquitoes were moving the night after such flights when no rain fell. Tshinaev (1945) reported that relative humid- ities of 100 percent depressed activities of mosquitoes.

Mosquitoes are usually more active during the night than during the day. Maximum nocturnal activity (based on attraction to light) takes place during the first three hours of darkness. Hourly trap records made nightly between June 14 and September 1 (summer conditions) showed that 64 percent of females were active during dark hours 1-3 and 36 percent were active during dark hours 4-6. After September 1 percentages for comparable times were 89 and 11 respectively. While

TABLE 21

EFFECT OF RAINFALL ON INCIDENCE OF FEMALE *Aedes vexans* COLLECTED IN A
STATIONARY INTERCEPTOR NEAR THE SANGAMON RIVER, PIATT COUNTY,
ILLINOIS, DURING JUNE AND JULY, 1960

Date	Rainfall (mm)	Mosquitoes collected (no.)
June 19-20	37	22
June 20-21	0	2
June 22-23	60	21
June 23-24	0	2
June 30-July 1	30	25
July 1-2	0	6
July 9-10	30	60
July 10-11	15	19
July 12-13	45	52
July 13-14	0	0

these percentages may vary from site to site, the contrast between incidence trapped early in the night and later will vary according to the thermal differential at a site. Love *et al.* (1963) reported that two-thirds of adults taken at light in Georgia were collected during dark hours 1-3. Knight and Henderson (1967) reported that two peaks of adult activity occurred in a city in Iowa. By means of a mobile interceptor moving over a fixed four-mile course along city streets during eight nights (June and July), 1,080 identifiable mosquitoes were caught; over half were *A. vexans*. Collections rose rapidly during the hour after sunset. A low order of activity was reported from then until shortly before sunrise, when a flurry of activity was noted. Over one-fourth of the trapped females were gravid, while only two contained fresh blood.

Stimuli that initiate exodus of *A. vexans* from an emergence site involve light, humidity, age of emergents, air movement, temperature, and possibly others. The first records of factors leading up to exodus of this species in central Illinois were made in 1958, when a brood emerged June 22. Exodus of part of that brood took place between 8:30 and 8:45 P.M. toward the west under a cloudless sky. The massive flight came the following evening while a light rain was falling and wind was gusting up to 10 km/hr.

The combined effects of stimulating factors is indicated in the departure of the two broods from the flood plain of the Sangamon River in 1960. (See also Haeger, 1960, and Nielsen, 1958, for stimuli initiating exodus of salt-marsh species.) Some diurnal lateral movement occurred, especially in wooded parts of the flood plain before exodus. Activity leading to departure from a grassy area on the north edge of the flood plain began as the sun neared the western horizon. Mosquitoes edged upward from their diurnal perches near the ground as humidity rose and light and temperature decreased. By the time of sunset, adults were on the inflorescences of the tall grasses (*Phalaris*) 100-125 cm above the ground. Brood II emerged in a concentrated period and departed within less than an hour after sunset on day of exodus. Brood I emerged over several days, and departure times took place during several nights. (See also Nielsen, 1958, and Provost, 1960, for comparison with *Aedes taeniorhynchus*.)

The phenology of two sequential broods of *A. vexans* that were produced on the flood plain of the Sangamon River in Piatt County, Illinois, are detailed herewith for 1960. (Note: Water level is indicated by +cm for height of flood, and by −cm for water level below flood level. Adult collections are based on number collected from ten samples, each 625 cm², taken within a rigid net forced into the grass.)

PHENOLOGY OF BROOD I.

June 16: rainfall 75+ mm over area of the watershed; water level, −15

June 17: water level, +37; hatching

June 18: water level, +30

June 19: water level, +23

June 20: water level, +7

June 21: water level, 0

June 22: water level, +37; larvae: instar 1 and 3

June 23: water level, +135; larvae: instar 1, 2, 4

June 24: water level, +195; larvae: instar 1-4

June 25: water level, +240; larvae: instar 1-4 and pupae

June 26: water level, +225; larvae: instar 2-4 and pupae

June 27: water level, +210; pupae abundant; 3 ♂, 0 ♀

June 28: water level, +180; pupae

June 29: water level, +90; pupae

June 30: water level, +60; pupae

July 1: water level, +45; rainfall: 10 mm

2 P.M.: pupae; adults: 74 ♂, 25 ♀ in marginal grass; some exodus

July 2: water level, +27; rainfall: 0

2 P.M.: pupae: 20-30/dip in woods; adults: 300 ♂, 357 ♀ (10% w/sperm); no biting

7:35 P.M.: males moving to tops of canary grass
8:25 P.M.: sunset; sky clear; wind 4.8 km/hr; temp 26°
male exodus begins; upward S @ 60° angle as singles and
clusters
8:35 P.M.: female exodus begins; upward S @ 60° angle as singles
and clusters; biting rate above flood plain: 9/min
8:40 P.M.: biting rate: 20/min decreasing to 6/min by 8:50
8:50 P.M.: exodus complete for day
July   3: water level: 0; rainfall: 0
2:00 P.M.: pupae in woods: few; adults: 57 ♂, 20 ♀
8:20 P.M.: sunset; sky clear; 22°; wind: NW @ 4.8 km/hr;
inversion layer @ 1.8 m
massive male exodus: vertical
8:30 P.M.: 18° @ 1.8 m to 19° @ 1.2 m and 22° in grass
8:42 P.M.: massive exodus (♂ & ♀), vertical with drift to NW
8:50 P.M.: massive exodus, upward W @ 20° angle; directional
interceptor trap above flood plain showed westward movement
8:51 P.M.: all flight ceased and air still
8:52 P.M.: slight riffle of air and mass exodus
9:00 P.M.: all flight ceased
July   4: water level, −5; 3 ♂ and 40 ♀
July   5: water level, −15; adults in grass: 4 ♂, 19 ♀
July   6: water level, −15; adults in grass: 7 ♂, 10 ♀
July   7: water level, −15; 1 ♂, 7 ♀

PHENOLOGY OF BROOD II

July 12: water level, −37
July 13: water level, +15; larvae: instar 1
July 14: water level, +110; larvae: instar 1 and 2
July 15: water level, +135; larvae: instar 1-3
July 16: water level, +110; larvae: 20-50/dip as instar 2-4
July 17: water level, +45
July 18: water level, 0; larvae: 30-40/dip as instar 3-4 w/balling
July 19: water level, −15
July 20: water level, residual pools; pupation begun
July 21: water level, residual pools; pupation complete
July 22: water level, residual pools; no adults emerged by 2:00 P.M.
July 23: water level, residual pools; all adults emerged by 2:00 P.M.
8:00 P.M.: adults low in grass; 112 ♂, 135 ♀ ; males: 50% rotated
8:45 P.M.: adults on heads of canary grass; no wind; exodus be-
ginning

July 24: rain in afternoon: 19 ♂ , 70 ♀
  8:03 P.M.: first adults to heads of grass
  8:07 P.M.: first exodus at 80° up and NW
  8:26 P.M.: exodus continues
  8:50 P.M.: exodus complete
July 25: no new adults in grass at 10:00 A.M.; exodus complete

The phenology of the two broods detailed above illustrates a variety of conditions that may affect the exodus of newly emerged mosquitoes. Brood II was the product of a single rapid rise of the river that produced larvae which became adults within about a twenty-four-hour interval beginning July 22 and ending July 23. Adults were densely stratified low in the wet grass during daylight on July 23. They began climbing the tall canary grass by sunset (8:00 P.M.). Exodus began at 8:45, and about half of the brood left the site that evening. The remaining mosquitoes departed between 8:00 and 8:50 the next evening. Adults rose from the grass nearly vertically (80°) and vanished in a northwesterly direction.

Conditions surrounding Brood I were complex. Adults comprising it were the product of two inundations of the flood plain (rise 1, June 17-18, hatched eggs at lower horizons; rise 2, June 22-25, hatched others at higher levels). Appearance of the first group of adults was further complicated by runoff from tributary channels flooding low places where larvae hatched first on parts of the flood plain before the river water left its channel. One such site was 1.6 km upstream from the place of record; emergence there began three days ahead of the site of record. The initial rise of the river reached maximum level of +37 cm; the second rise produced a level of +240 cm and covered the entire flood plain to maximum depth for the season. As a result of sequential rises in water level over several days, larvae hatched at the site of record over a five-day span and in the vicinity over about eight days. The water was cool (20°) under the canopy and warm (25°) when it was sunlit in the grassy areas. Emergence was spread over several days from June 30 to July 5. Consequently, exodus was spread in time (July 1-8). Most exodus occurred on July 2 and 3, yet some non-inseminated females were still present on July 11.

Males of Brood I congregated in the trees over the wooded part of the emergence site in the dusk and after dark. They were present and flying in such numbers that an audible humming sound could be heard by a person on the ground. Such was not the case with Brood II, because all adults left the flood plain completely. (Smith [1903] reported a similar situation with a summer brood of *A. sollicitans* in New Jersey. In this instance males emerged "in clouds." Two days later "just at dusk a peculiar humming noise seemed to fill the air." It was caused by clouds of mosquitoes 1.8 to 6 m above the marsh.)

Movement to a light trap 100 m north of the emergence site for Brood I showed the first rise in incidence took place between 9:00 and 11:00 P.M. on June 28. During the interval 10 ♂ and 12 ♀ (newly emerged) arrived. The next night (July 29) 29 ♂ and 6 ♀ arrived. A total of 29 ♂ and 94 ♀ arrived during the two-hour interval of June 30. For July 1-4 the numbers were 18 ♂ and 16 ♀ ; 30 ♂, 27 ♀ ; 57 ♂, 85 ♀ ; 17 ♂, 9 ♀ respectively. Movement in the area continued between exodus of Broods I and II.

The chronology of events leading to exodus of a mid-summer brood of *A. vexans* in 1962 is shown in Table 22. Emergence was concen-

TABLE 22

PHENOLOGY AND ACTIVITY OF AN EMERGENT BROOD OF *A. vexans* FROM FLOOD PLAIN IN CENTRAL ILLINOIS IN LATE JULY, 1962

| Date | Hour | Wind | | Temp at 120 cm | Sun angle | Sky | Activity |
		Di-rec-tion	Speed (m/min)				
July 19	2:00 P.M.	..	..	..	..	..	Emergence complete
July 20	6:00 P.M.	WSW	83	28.5°	15°	veiled	Perching: 0-25 cm
	6:45 P.M.	WSW	67	28°	10°	veiled	Perching: ♂ 45 cm; ♀ 0-25 cm
	7:15 P.M.	WSW	106	27°	5°	clear	Launch: windward; direction: SSE
	7:42 P.M.	WSW	28	25.5°	0°	clear	♂: in trees and partial rotation
	8:00 P.M.	0	0	24.5°	..	clear	Direction: ♂, ♀ S to trees to perch
	8:15 P.M.	0	0	24.5°	..	clear	No exodus
July 21	7:15 P.M.	N	45	28.5°	5°	clear	Perching in grass
	7:45 P.M.	0	0	27.5°	1°	clear	Creeping flight and ♂ up 60° S to canopy
	8:15 P.M.	0	0	25°	..	clear	♂, ♀ up 60° S to canopy
	8:45 P.M.	0	0	23°	..	clear	Exodus upward through canopy
	9:00 P.M.	0	0	21°	..	clear	Residue of adults perched in marginal grass

trated in time and completed on July 19. Conditions provided dry air, high temperature, clear sky, and little or no wind during the time of exodus on July 20 and 21. Exodus was desultory and incomplete during the interval. Greatest tendency was for the brood to move laterally near the ground. Part of the brood dispersed as the mass did in 1960 (see above). Those that left July 20 did so during the time the wind was blowing (7-8:00 p.m.). Launch was windward, but departure was downwind toward a nearby sylvan canopy. Mosquitoes were abundant on leaves of walnut trees in the edge of the emergence site. Three to ten adults perched on each leaflet at least within 5 m of ground level. No humming sounds were heard, as was the case in 1960. Creeping flight to marginal grass and shrubs was the rule. Residue of females at the site were inseminated, and presumably they did not vacate the area.

A device was rigged to appraise the effect of changes in light, temperature, and humidity on initiation of the flight response by a laboratory population of unmated, unfed adults during several hours after emergence. The device consisted of a chamber with nonreflective black walls inside a room with overhead fluorescent lighting which was manipulated in intensity by a rheostat. Bright light had an intensity of 6.5 on the meter used, and dim light had an intensity of 1. Humidity was regulated by a hygrostat operating an aerosol-emitting humidifier behind a baffle. A silken thread was cemented to the mesonotum of each mosquito. The thread extended upward into a capillary tube filled with oil (SAE-50). Feet of the tethered mosquitoes rested on a black plate. They were free to take flight on their own initiative.

Mosquitoes were held for four to eighteen hours after emergence under varying conditions of light, humidity, and temperature. They were then exposed for thirty minutes or until they took wing, whichever came first. The lighting of dusk was simulated by decreasing intensity of overhead light from bright to dim with a rheostat. Dawn was simulated by raising the light intensity from 1 to 6.5 on the meter. Shifts of temperature and humidity under simulated dusk are indicated in Tables 23-28. Much of the recording of results were made by Dr. R. A. Brust, a research assistant at the time.

Initiation of flight was stimulated earlier and most consistently under conditions of simulated dusk when humidity was high (Tables 23 and 25). Simulated dawn and high humidity activated mosquitoes somewhat later (Tables 24 and 26). Low humidity definitely inhibited the flight response (Tables 27 and 28). Mosquitoes held at simulated dusk and 28° throughout began to fly at an earlier age; all held at that temperature began to fly by the time they were ten to twelve

TABLE 23

EFFECTS OF DECREASED LIGHT ON INITIATION OF FLIGHT RESPONSE BY UNFED,
UNMATED *Aedes vexans* HELD IN BRIGHT LIGHT AT 28° AND 90% RH

Age of mosquito (hrs.)	Mosquitoes treated (no.)	Mosquitoes flying within 30 min in dim light	
		No.	%
4	18	4	22
6	18	9	50
8	18	13	73
10	18	16	89
12	18	18	100
18	18	18	100

TABLE 24

EFFECTS OF INCREASING LIGHT ON INITIATION OF FLIGHT RESPONSE BY UNFED,
UNMATED *Aedes vexans* HELD IN DIM LIGHT AT 28° AND 90% RH

Age of mosquito (hrs.)	Mosquitoes treated (no.)	Mosquitoes flying within 30 min in bright light	
		No.	%
4	..	..	..
6	18	3	17
8	18	8	44
10	18	9	50
12	18	12	67
18	18	15	83

TABLE 25

EFFECTS OF DIMINISHED LIGHT AND TEMPERATURE (28° TO 20°) ON INITIATION OF
FLIGHT RESPONSE BY UNFED, UNMATED *Aedes vexans* HELD AT 90% RH

Age of mosquito (hrs.)	Mosquitoes treated (no.)	Mosquitoes flying within 30 min in dim light	
		No.	%
4	18	0	0
6	30	3	10
8	30	13	43
10	18	10	56
12	30	27	90
18	30	30	100

TABLE 26

EFFECTS OF INCREASED LIGHT AND DECREASED TEMPERATURE (28° TO 20°) ON
INITIATION OF FLIGHT RESPONSE BY UNFED, UNMATED
Aedes vexans HELD AT 90% RH

Age of mosquito (hrs.)	Mosquitoes treated (no.)	Mosquitoes flying within 30 min in bright light	
		No.	%
4	18	3	17
6	18	4	22
8	18	6	33
10	18	10	55
12	18	11	61
18	18	10	55

TABLE 27

EFFECTS OF DECREASED LIGHT ON INITIATION OF FLIGHT RESPONSE BY UNFED,
UNMATED *Aedes vexans* HELD IN BRIGHT LIGHT AT 28° AND 40% RH

Age of mosquito (hrs.)	Mosquitoes treated (no.)	Mosquitoes flying within 30 min in dim light	
		No.	%
4	18	2	11
6	18	5	28
8	18	7	39
10	18	10	56
12	18	10	56
18	18	11	62

TABLE 28

EFFECTS OF INCREASED LIGHT ON INITIATION OF FLIGHT RESPONSE BY UNFED,
UNMATED *Aedes vexans* HELD AT 20° AND 40% RH

Age of mosquito (hrs.)	Mosquitoes treated (no.)	Mosquitoes flying within 30 min in bright light	
		No.	%
4	18	0	0
6	18	0	0
8	18	4	22
10	18	8	44
12	18	11	61
18	18	12	67

hours old (Table 23). Even when the dusk temperature was dropped to 20° after prior 28°, all of the mosquitoes were airborne within thirty minutes at twelve to eighteen hours of age (Table 24). No more than two-thirds of the mosquitoes became airborne within thirty minutes when held at 40 percent RH regardless of age (Tables 27, 28).

Feeding. Adults feed primarily on fluids such as nectar derived from plants, and the females feed additionally on blood of birds and mammals. An early report, Knab (1907), gives an account of large numbers of both sexes feeding on flowers of *Solidago.* Philip (1943) noted that females of all ages feed on nectar of this group in the northern Great Plains of USA-48. In British Columbia, Hearle (1926) found adults of both sexes feeding on flowers of spiraea.

Blood seems to be required by females for development of eggs, and no records of autogeny have been found. Hundreds of mosquitoes have been reared in this laboratory, but none has oviposited without a prior blood meal. Reeves and Hammon (1944) reported that a population in the Yakima Valley of Washington had fed on horse (20%), cow (56%), fowl (4%), unknown (20%). An Hawaiian population examined by Tempelis *et al.* (1970) had fed on horse (37%), cow (51%), dog (5%), man (2%), and other (5%). Nearly all of an urban collection of *A. vexans nocturnus* in Cambodia had fed on bovine blood (92%), and 8% on man (Jolly *et al.*, 1970). Reeves and Rudnik (1951) found man, cow, and dog to be hosts of a population on Guam. A Florida population taken from an aggregation site a mile or more from available bovines had fed largely (81%) on cattle blood, according to Edman and Bidlingmayer (1969). Shemanchuk (1969) reported that females fed on avian hosts readily in Alberta, Canada. A Kansas population had fed as follows: cattle, 62%; man, 7%; sheep, 12%; remainder on rabbit, pig, horse, bird, and dog (Edman and Downe, 1964).

Females may feed on blood during both day and night, but most active feeding is early in the evening at summer temperatures. Wright and Knight (1966) observed that feeding was concentrated largely in an interval from near sunset until some three hours after sunset on a wooded flood plain in Iowa. Roberts *et al.* (1969) noted that 97 percent fed between 5:00 P.M. and 7:00 A.M. in Mississippi. Sasa (1964) noted a like habit for a Japanese population. The exact time of day when feeding is maximal may vary according to whether the period after emergence is one when nights are cold or when days are hot and dry. Feeding is minimal or absent at night in central Illinois when the air temperature near a collector is below 19°, but between 19° and 25° the rate of feeding at night increases with temperature in proportion to abundance of the population. During the day temperatures of

25° or above may inhibit feeding in sunshine when humidity is low. On one occasion feeding was observed in the evening when air temperature was 17°. When *A. vexans* emerges early in the season, aggregations of biting populations are near larval sources and are most aggressive in the afternoon. The early brood feeds avidly on man under conditions where prolonged cold weather inhibits dispersal after emergence. During the 6-7:00 P.M. interval on May 15, 1954, 148 *A. vexans* were attracted to the forearm of a collector. Temperature since emergence had remained below 15° for a week and had risen for the first time to 20°+ on the day of collection. Thompson and Dicke (1965) observed that a summer brood fed more actively in grasslands than in a forest (ratio 3:1) in Wisconsin. Peak periods of feeding were 8:00 P.M. and 8:00 A.M. These observations may well reflect times of optimum thermal conditions.

Aedes vexans is not as readily attracted to man in sylvan sites during the day as is *Aedes stimulans*, for instance. The reverse is true in the early evening (7:00 to 7:30). From May 14 to June 8, 1960, a man with an arm exposed collected 280 of the former and 37 of the latter during a season's cumulative collecting time of 280 minutes. *Aedes vexans* was as readily attracted to a stationary collector as to one walking, where *A. stimulans* chose a mobile host over a sitting one three times as often.

Relative numbers of *Aedes vexans* and *stimulans* attracted to man and to light are reversed, according to collections recorded by Porter and Gojmerac (1969) in a sylvan site in Wisconsin. Fifty to 70 percent of light-trap catches were *A. vexans* and 8-19 percent were *stimulans*, while 1 to 9 percent of the mosquitoes biting man were of the former and 50 to 80 percent of the collections were of the latter species.

The amount of blood ingested at the time of a first feeding was determined to be 2.0 mgm on the average in a population of 120 females examined by Stage and Yates (1936). Woodard and Chapman (1965) reported the average increase in weight by engorgement to be 4.7 mgm for fifty mosquitoes.

Feeding on blood normally follows copulation under natural conditions. No non-inseminated females have been collected while biting in the field, although several hundred have been dissected and spermathecae ruptured. Feeding without prior copulation in the field might occur in the early spring when dispersal is erratic. Insemination is not required to initiate feeding in laboratory-reared populations. At 25° a female may engorge on blood after forty-eight hours beyond emergence. However, feeding may be sporadic during the first week at this temperature. Engorgement of females mated in the laboratory takes place within six hours after mating, while unmated ones engorge erratically and less avidly.

In east-central Illinois the season when mosquitoes of any species may be expected to bite begins as sporadic attacks by a few over-wintering species (*Culiseta*, *Anopheles*, and *Culex*) in April and continues well into October. During 1956 between April 26 and early October, 72 collections representing 1,250 minutes of exposure of a forearm were made in a sylvan site near Urbana, Illinois (Table 29). A total of 1,029 mosquitoes representing 11 species were attracted. *Culiseta inornata*, *Anopheles punctipennis*, and *A. quadrimaculatus* Say came during late April and early May. *Aedes stimulans* appeared in large numbers in May. *Aedes vexans* and *A. trivittatus* appeared sporadically during the last half of May and were most numerous during the summer. Thirty-four percent of the population attracted throughout the season was *Aedes vexans*. During September 82 percent of the population was this species. The near absence of biting mosquitoes in the interval May 20-28 does not reflect an absence of them from the site. This was an interval when temperatures early in the evening were too low for their activity. They were present in the herb layer but were not host-oriented. Rees (1943) noted that the average annoyance time for a brood that has invaded a city is ten days.

Copulation. Nothing is known about copulation of this mosquito in the wild. Presumably it takes place at the time of exodus and in the air over the site, as has been reported for *A. taeniorhynchus* (Haeger, 1960). Males exit a few minutes before the females and can be heard overhead during the exit flight of females. Females are not inseminated prior to exodus flights, but they are inseminated when trapped within 100 m, as local records show. Some inseminated ones return to the emergence site immediately. Since males accompany females in mass flight, possibly insemination may occur elsewhere than in the air above emergence sites. On one occasion copulation was noted early in the morning in tall grass in an urban arrival site while dew was present.

Aedes vexans does not copulate normally in cages, and none will copulate in small cages. Dashkina and Tsarichkova (1965) observed copulation in cages 200 x 100 x 100 cm at 80-90 percent RH during simulated dawn and dusk periods of forty minutes each. All efforts to provide cage conditions that encourage free copulation of a local population have been unproductive of inseminated females.

Laboratory culture of the nearctic population is made routinely possible by induced copulation. Adults varying in age from five to fourteen days were routinely used for mating stock for maintenance

TABLE 29

RELATIVE SEASONAL INCIDENCE OF *Aedes vexans* AND OTHER MOSQUITOES BITING MAN DURING TWO HOURS PRIOR TO DUSK IN A SYLVAN SITE NEAR A PROLIFIC SOURCE OF FLOODWATER MOSQUITOES, CHAMPAIGN COUNTY, ILLINOIS

Collecting interval	Days collecting	Minutes collecting	Mosquitoes collected per 100 min										
			Aedes vexans		*Aedes trivittatus*		*Aedes stimulans*		*Psorophora horrida*		Other[a]	
			No.	%	No.	%	No.	%	No.	%	No.	%
April 26-May 18	17	200	2	3	0	0	55	89	0	0	5	8
May 20-May 28	8	170	28	28	19	19	60	60	0	0	3	3
May 29-June 9	8	170	0	0	0	0	2	100	0	0	0	0
June 11-June 27	7	90	22	16	92	68	6	5	12	9	2	2
June 30-July 22	9	105	47	26	110	61	6	3	13	7	6	3
July 24-Aug. 23	11	205	7	9	58	77	0	0	4	5	6	8
Aug. 29-Sept. 9	6	160	51	63	20	25	0	0	1	1	9	11
Sept. 13-Oct. 1	6	150	83	100	0	0	0	0	0	0	0	0

[a] Includes *Culiseta inornata*, *Aedes sticticus*, *A. triseriatus*, *Psorophora ferox*, *Culex salinarius*, *Anopheles punctipennis*, and *A. quadrimaculatus*.

of laboratory strains. Manipulation of adults during the mechanics of copulation was according to the technique reported by Horsfall and Taylor (1967). (See *Colonization.*)

Induced copulation of females of *A. vexans* is readily achieved by active, decapitated males and females immobilized by anesthesia. The influence of anesthetics on rates of insemination was investigated by exposing virgin females to atmospheres of nitrogen, carbon dioxide, or chloroform. Single decapitated males were mated with virgin females for three repetitive copulations, and rates of insemination were determined by dissection and examination of spermathecae for presence of spermatozoa.

The efficiency of insemination of virgin females of *A. vexans* varied according to the anesthetic used prior to inducing copulation and by the sequence of mating with males (Table 30). (See also Fowler, 1972.) When virgin males were induced to copulate with virgin females, 90 percent of the females exposed to nitrogen and carbon dioxide had one or more thecae filled with spermatozoa, in contrast to a rate of insemination of 50 percent for females exposed to chloroform. Spermatozoa were present in all three thecae in 60 percent of the females immobilized by nitrogen and 20 percent of the females treated with carbon dioxide. Copulation of once-mated males with virgin females resulted in consistently high rates of insemination for all. Rates of insemination of 90, 70, and 80 percent were obtained for females anesthetized with nitrogen, carbon dioxide, and chloroform, respectively. For this series 40 percent of the females exposed to nitrogen and 10 percent of those exposed to chloroform had sperm in all thecae. On the third copulation of males with virgin females, rates of insemination remained high for the nitrogen group but diminished for the other two. Eighty percent of the females exposed to nitrogen were inseminated, in comparison to 50 percent for specimens in the carbon dioxide group and 40 percent for females in the chloroform group. Thirty percent of the females exposed to nitrogen had sperm cells in all thecae, while females in the other two groups had no more than two thecae positive for sperm.

Copulation may occur without successful insemination of females. The frequency of occurrence is lowered if only males which exhibit a strong mating response following decapitation are used for mating. By using vigorous males, a high percentage of females may be successfully inseminated up through the third repetitive copulation. During this study copulation without insemination was especially noted when females were held by males for coital periods of less than six seconds or for periods exceeding thirty seconds.

TABLE 30

EFFECT OF THREE DIFFERENT ANESTHETICS ON INSEMINATION OF 90 VIRGIN
FEMALES OF *Aedes vexans* INDUCED TO COPULATE 1-3 TIMES

Copu-lation times (no.)	Fe-males copu-lated (no.)	Number of females inseminated									Total
		Theca 1 w/sperm			Thecae 1, 2 w/sperm			Thecae 1, 2, 3 w/sperm			
		N_2	CO_2	$CHCl_3$	N_2	CO_2	$CHCl_3$	N_2	CO_2	$CHCl_3$	
1	30	2	0	2	1	7	3	6	2	0	23
2	30	0	0	3	5	7	4	4	0	1	24
3	30	1	0	3	4	5	1	3	0	0	17
Total	90	3	0	8	10	19	8	13	2	1	64

The time required for coitus during induced copulation of virgin
females by single decapitated males is presented in Table 31. Time of
contact was measured from the time of emplacement of the aedeagus
until its retraction. Termination of coitus was characterized by a rapid
upward thrust of the dististyles and simultaneous release of females.
The coital time periods presented in Table 31 were obtained during an
experiment on the effect of anesthetics on rates of insemination. Males
were successively induced to copulate with three females, and a total
of ninety females were mated with thirty males.

Induced copulation caused biting activity to vary according to the
anesthetic used to immobilize females prior to mating (Table 32).
Human blood was offered at time intervals of 1, 6, 24, and 36 hours
after mating. One hour after mating, more than 50 percent of the females immobilized with nitrogen and carbon dioxide fed on human
blood, in contrast to only 5 percent of the females in the chloroform
group. After a post-mating interval of six hours, 90 percent of the
females treated with nitrogen and carbon dioxide had fed on blood,
while only 35 percent of the females anesthetized with chloroform had
taken a blood meal. This general pattern persisted throughout the
study period. After a post-mating interval of twenty-four hours, almost
all of the females exposed to nitrogen and carbon dioxide had taken
a blood meal, while a time interval of thirty-six hours was required
for 75 percent of the females treated with chloroform to take blood.
Similar results were obtained when biting activity was studied with
virgin females.

Maturation. Aedes vexans, like most aedine mosquitoes, are anautogenous — i.e., they require that blood be ingested before maturation of

TABLE 31

COITAL TIME PERIODS FOR VIRGIN FEMALES OF *Aedes vexans* INDUCED TO COPULATE SUCCESSIVELY WITH SINGLE DECAPITATED MALES AFTER EXPOSURE TO DIFFERENT ANESTHETICS

Anesthetic used	Males used (no.)	Females used (no.)	Coital time in seconds								
			Copulation 1			Copulation 2			Copulation 3		
			Min.	Max.	Mean & S.E.	Min.	Max.	Mean & S.E.	Min.	Max.	Mean & S.E.
Nitrogen	10	30	7	18	11.9 ± 1.0	7	18	11.3 ± 1.2	6	27	12.0 ± 1.8
Carbon dioxide	10	30	8	15	11.2 ± 0.6	7	14	11.7 ± 0.7	9	24	12.9 ± 1.5
Chloroform	10	30	11	16	13.1 ± 0.6	9	19	13.7 ± 1.4	10	18	14.3 ± 1.1

TABLE 32

EFFECT OF DIFFERENT ANESTHETICS USED DURING COPULATION ON DELAY
OF FEEDING OF MATED FEMALES OF *Aedes vexans*

Post-mating period (hrs.)	Females feeding on human blood					
	Nitrogen		Carbon dioxide		Chloroform	
	No.	%	No.	%	No.	%
1	25	62	31	77	2	5
6	11	28	5	13	12	30
24	2	5	3	7	8	20
36	2	5	0	0	8	20
Total	40	100	39	97	30	75

eggs progresses. Nothing is known about the minimum amount required, although blood in the gut is required to activate maturation as well as to nourish eggs. Lea (1970) explains activation of maturation on secretion from the corpora allata and one from the medial neurosecretory cells of the endocrine system of mosquitoes. Presumably, autogenous females secrete both hormones soon after emergence, and those from the corpora allata may be activated before engorgement on blood. In the case of anautogenous *A. taeniorhynchus*, at least, a blood meal is required to release an essential hormone produced by the medial neurosecretory system that is stockpiled in the corpus cardiacum.

A wild population in Wisconsin showed a parous state in 30-40 percent of the females about one week after peak emergence for each brood based on collections at carbon dioxide–baited traps (Morris and De Foliart, 1971). This trapping procedure is biased for blood-seeking females which for the most part are seeking their first blood meals.

Oviposition. Aedes vexans oviposits on soil subject to periodic inundation. Soil exposed above the water table after prolonged inundation is particularly attractive to gravid mosquitoes. When the exposed soil is sandy, covered by plant detritus (logs, stems, and leaves) or high in humus, it receives large numbers of eggs. Old lake beds with firm soil, margins of flood plains of rivers, and grassy swales are ecological units attractive to ovipositing adults (see *Egg: Focal distribution*).

In the spring of 1954 the rate of deposition of eggs for the spring brood of *Aedes vexans* was observed at a prolific oviposition site near Urbana, Illinois. Eggs deposited the prior season hatched during two successive rises of the water level in mid-April. Adults began emerging April 28, and by the second week in May they began to return to oviposit. Four series of ten samples were taken (as shown in Table 33) during the interval May 10-28. Very few new eggs had been deposited by the first day of sampling. By May 17 the incidence of new eggs had risen vastly over the week before, and oviposition had stopped by this date.

The second brood of *Aedes vexans* emerged from this station June 8, following a flooding June 1. All eggs of the previous brood hatched during the flooding. Sampling began June 14 while lower horizons were still under water. About half of the eggs had been deposited by June 14, and the remainder were in place by June 17-24. This brood showed a sharp decline in incidence of new eggs at the horizon of sampling by July 1 (Table 33).

Hearle (1926) reported in his work with a valley population in British Columbia that eggs were fully developed in the ovaries early in July, and oviposition of the brood continued until the end of August. A large proportion of the females taken by net at oviposition sites after mid-July were fully gravid. All eggs in the ovaries were in the same stage of development. The number of eggs varied from 108 to

TABLE 33

RATE OF DEPOSITION OF EGGS BY TWO SUCCESSIVE BROODS OF *Aedes vexans* IN A WOODLAND POOL NEAR URBANA, ILLINOIS, 1954

Date of collection	Area sampled (cm^2)	New eggs per 100 cm^2
May 7	2,250	0
May 10	2,250	<1
May 14	2,250	7
May 17	2,250	39
May 21	2,250	24
May 26	2,250	10
May 28	2,250	5
June 1	0[a]	. .
June 14	670	15
June 17	1,125	29
June 21	1,125	29
June 24	1,125	32
July 1	1,125	11
July 10	1,125	0

[a] Site submerged.
Brood I emerged by April 28.
Brood II emerged June 8.

182 (average, 132). When voided, the eggs were ejected very rapidly, each following the other at intervals of a few seconds.

The population in central Illinois produces fewer eggs on the average than does that of British Columbia, but individual females have deposited up to 374 eggs during a life span. Thompson and Dicke (1965) observed that wild-caught adults deposited from 4 to 205 eggs. Oviposition habits of 132 wild-caught females caged individually were observed in this laboratory. Eighty-nine oviposited once only, fourteen oviposited twice, eighteen oviposited three times, and one oviposited four times. Records kept on ingestion in relation to oviposition (Table 34) indicate that wild-caught females take more blood when held at 25° than when held at 20° but that the average number of eggs deposited is not appreciably different at the two temperatures. One blood meal is adequate for oviposition, as was noted here and by Mitchell (1907). In the laboratory many females die the day of oviposition, but on the average they may live much longer. Thompson and Dicke (1965) noted 85 percent mortality on the day of oviposition.

The oviposition capability of reared and manually copulated individual females was observed in this laboratory. The number of eggs per female ranged from 97 to 374 (Table 35). Three of the females deposited one egg mass each, four deposited two masses each, and two deposited four times. One blood meal was adequate for deposition of over 140 eggs. Five of the females took only one blood meal, and one of them took seven meals. Longevity ranged from seven to eighteen days under the conditions imposed.

The capability of an Alabama population was more variable (Breeland and Pickard, 1964). Eggs per female varied from 22 to 546, with

TABLE 34

Oviposition of Wild-caught Females Collected in the Spring

Date of collection	Number of females	Laboratory temperature (°C)	Number of blood meals (average)	Eggs deposited per female (average)
April 5 + 6	13	25	4.4	88
April 7	12	25	3.1	85
April 8	15	25	2.8	95
April 9	15	25	3.1	92
April 11	17	25	3.2	85
April 20	19	25	4.2	123
May 19	18	20	2.2	98

TABLE 35

Habits of a Reared Population of *Aedes vexans* Manually Copulated Two Days after Emergence Held in Cages at 25° and in Bright Light

| Mosquito | Activity of each mosquito | | | Longevity (days) |
| | Blood meals | Oviposition effort | | |
		Times	Viable eggs	
1	7	4	374	18
2	1	1	141	11
3	4	4	299	16
4	1	1	102	7
5	1	2	97	9
6	3	2	267	12
7	1	1	149	9
8	3	2	265	11
9	1	2	109	10

the number of oviposition efforts ranging up to 12 and the number of eggs deposited at one time ranging up to 226.

At 25° females will oviposit about six days after the first blood meal, even though two or more feedings may take place. After a blood meal the mosquitoes are lethargic for two or three days; then they may become more active for a time. During the period of activity they void black, tarry feces. Oviposition begins one or two days after voiding the black residue.

Excretion. Females defecate frequently, but nothing is known of the specifics of the content of the excreta. At the time of blood feeding the female passes clear droplets of liquid which are forcibly ejected. After engorgement on honey or blood, defecation is frequent. The floor below the resting site of caged mosquitoes may be coated with coalesced droplets. For a few days after a blood meal dried fecal droplets have a light tan color. One or two days before oviposition the color is nearly black. Presumably the black, tarry material is partially digested blood.

Secretion. Very little is known about the specifics of secretions for this species (Allen and West, 1962). Metcalf (1945) stated that saliva contained neither anticoagulin nor agglutinin for human blood. Oral secretions producing sensitivity reactions in man and other animals have been noted by Benson (1936) and McKiel and West (1961). Saliva does have an urticating fraction and histiminic activator as judged by human reactions to salivation. Nothing is known of secretions affecting digestion, maturation, insemination, and oviposition. The female has a tiny accessory gland opening into the atrium of the genital tract. Nothing is known of its function in *Aedes vexans*. The male has two large accessory glands opening laterally into the ejaculatory canal. These secrete a fluid in which spermatozoa are transferred during copulation.

Aggregation. This mosquito forms aggregations at least four times during adult life. The first of these is a post-emergence massing of both males and females in low vegetation marginal to emergence sites. Where emergence is from woodland pools, adults move to the marginal herbaceous layer. Where emergence is from grassy swales, adults move high in the grass or to marginal herbs and grass. On flood plains which are inundated, adults move to vegetation above the water line. They perch in the lower levels of grass or herbs at various heights above ground according to humidity. Vertical spread may be wider after dark. (See also *Adult: Focal distribution.*) Bresslau (1917) noted that broods emerging early in the spring remain longer near emergence sites. Post-emergence massing may last only a day or even less when climatic factors are most favorable to mass exodus.

Post-dispersal, diurnal massing is variably concentrated in shrubbery, tall herbs, and tall grass in pastures, and in towns and cities near sources of blood. Adults avoid short grass and areas exposed to wind in this phase. They cling to lower or leeward sides of herbaceous leaves or to stems near the ground where humidity is suitable in dry weather. Aggregations are more stable in dry than humid air. Feeding and maturation are activities of this phase.

Gravid females aggregate in vegetation near oviposition sites. While the masses may be small and more dispersed in this phase, they are nonetheless real. In central Illinois the vegetation of choice of woodland and flood-plain sites is stinging nettle, an herb that forms an open canopy 50-90 cm above the ground and provides a fairly unobstructed zone below for movement. Aggregations of gravid adults are evanescent and poorly defined.

The tendency toward nocturnal gathering at light is well known and has been exploited by abatement districts to appraise population levels with lighted traps. Many factors modify reliability of such traps as attractants, the major one of which is proximity and position of competitive lights. The greater the isolation of the attracting light, the more the concentration, as a rule. Breyev (1963) indicates that the population in the Volga delta (USSR) varied seasonally in its attraction to light from a high of 92 percent in late June to 32 percent in early August. While some of this decline could be a decrease in real population level, more likely weather limits movement toward light. Before aggregation at light is accepted as an indication of relative incidence, more study is needed of bias in orientation toward light.

At the time of exodus flights males form swarming aggregations in the canopy over or near emergence sites. Smith (1903) first reported massive swarming aggregations several meters above the ground. Dyar (1917) noted males swarming in the canopy over river bottoms. Small swarms were noted by Dyar in open glades, especially on the darker sides of bushes. Hearle (1926) reported post-emergence swarms of males at dusk over meadows, fields, and open glades in southwestern Canada. Massive blanket swarms have been noted on two occasions in Illinois, and both occurred above flood plains of rivers after sunset. Hearle considered swarming a prelude to copulation; such would seem to be the case when blanket swarms form in the canopy over flood plains. Tiny swarms (4 to 20 males) have been seen on the downwind side of a warm automobile in an alfalfa field near an emergence site in Illinois. No females were seen near such swarms. Certainly mating is not restricted to massive swarms, as copulating pairs have been seen in grasslands in urban areas after primary dispersion.

Latency. In the usual manner for aedine mosquitoes, this one has no facility for prolonged imaginal survival, as do species of *Culex, Culiseta,* and *Anopheles.* Short-term extensions of longevity may occur when cold weather forces reduced activity or a shortage of hosts defers ingestion of blood. No accumulation of stored reserves for prolonged survival occurs.

Longevity. Adult life of three weeks in summer and six weeks in early spring are about the limits of normal expectancy. Males usually die earlier than females. Males have survived in the laboratory and have mated twenty-one days after emergence when held at 21°. Females fed only diluted honey lived six weeks even at 25°, but few were alive more than six weeks. Costello and Brust (1971) noted that females lived seventy-six days at 21° or eighty-two days at 13°, while males lived no more than sixty-two days. Maximum longevity of females was uniformly higher at 80 percent RH than at lower levels. Stage *et al.* (1937) recorded life of fifty-five days for one female outdoors that was stained in a post-emergence aggregation and retrieved by trapping. Breeland and Pickard (1964) found longevity of wild-caught young females varied from six to forty-eight days when brought into the laboratory. (See *Adult: Feeding* section for longevity of blood-fed females.)

Antagonists. Two mermithid parasites have been seen attacking this mosquito. Steiner (1924) noted infection rates as high as 80 percent by *Paramermis canadensis* Steiner. Two to six worms might be present in one female mosquito. This nematode is found in the hemocoel and is a significant regulatory parasite in southern Canada. Trpis *et al.* (1968) noted this and an *Agamomermis* sp. in British Columbia. All adults collected at one site were infected. The other and sometimes common parasite is a water mite (species unknown) that attaches to membranes of the integument. Sometimes six or more mites are attached, but one or two are usual. Infested females do not deposit viable eggs. Mites are found more commonly on mosquitoes that remain near emergence sites than on those that emigrate for blood meals. Because broods that emerge early in the season tend to remain in the vicinity, mites are more common on *Aedes vexans* early in the year in central Illinois. Jalil and Mitchell (1972) found thyasid mites on the coxae of females.

Reservoir. Mosquitoes are involved as propagative and cyclopropagative hosts for microbes that attack man and other vertebrates. In this capacity the mosquitoes are part of the complex of hosts (reservoir) that maintains and extends the focality of the microbes. Among the groups of organisms having *Aedes vexans* in their reservoirs are viral agents and filariids.

The viral agents to which tissues of *A. vexans* are hospitable are shown in Table 36. The species acquires viruses that circulate with the blood in a number of species of birds; it has been shown to support a number of viruses, and evidence from laboratory and wild state indicates epidemiological importance of the species.

Tahyna virus from central Europe infects this mosquito in the laboratory and in the field, and *A. vexans* has been shown to act as vector in the laboratory. Virus at 3.37 log $LD_{50}/0.03$ ml may infect mosquitoes and appear in head and thorax after seven days and persist at least to day 26 (Danielova, 1962). Infection rates of 88 percent and transmission rates up to 90 percent have been established for those fed in the laboratory (Danielova, 1966). Transmission by biting may take place after four days (Simkova *et al.*, 1960). Infected mosquitoes (newly engorged) have been collected in cattle sheds and stables (Bardos and Danielova, 1959). The virus persisted in *Aedes vexans* for thirty days in values of 10^{-3} and 10^{-4} LD_{50}. Titers of the virus decrease on days 3 and 4 after engorgement and increase from day 5 onward (*idem*). Rate of infection of pigs may be 18 percent, but pigs seem to be terminal hosts, as viremic ones do not impart virus to *A. vexans* at least (Bardos and Jakubik, 1961). Malkova *et al.* (1965) obtained eleven isolations from 6,668 females trapped in Czechoslovakia in 1963-64. Aspöck and Kunz (1967) considered the steppe as providing conditions for basic foci, and marshy forests in mid-July to mid-September as expansion foci, where *A. vexans* abounds in central Europe.

The function of *A. vexans* in the reservoir of the virus of California encephalitis is not clear, but evidence indicating its importance is mounting. Isolations from a wild population of mixed *A. vexans* and *A. canadensis* (McLean *et al.*, 1970) have been made in British Columbia, where the species abounds. Ackerman *et al.* (1970) obtained fifty-four strains in Germany from 30,000 mosquitoes (largely *A. vexans*). Snowshoe hare and Columbian ground squirrel, normal hosts of *A. vexans*, show high incidences of antibodies.

Coincidence of this mosquito and the viruses of eastern (EEE) and western (WEE) equine encephalitis in feral foci is indicative of significance in their maintenance. This mosquito has been suspect as a vector of virus:EEE since 1935, when it was shown by Ten Broeck and Merrill (1935) to be a laboratory vector. Wallis *et al.* (1960) first isolated the virus from *A. vexans* during an epidemic among pheasants in Connecticut in 1959. Among 148 adults in ten pools, virus appeared once in a pool of thirteen females, none of which contained new blood. Sudia *et al.* (1968) isolated the virus once from pools of 792 females near the Gulf Coast of Alabama. Some fifteen transient and resident birds of upland woods where *A. vexans* abound were positive for the virus (Stamm, 1968). According to Chamberlain (1958) woodlands of freshwater swamps in southern USA-48 are foci for virus:EEE. Hayes (1962) presumed that this species was an inland zoonotic vector during an outbreak in New Jersey in 1959. Outbreaks in man and horses have generally originated in the vicinity of such sites. In the laboratory a high percentage (63%) of *A. vexans* became infected (*idem*). Davis (1940) used this mosquito to transmit virus:EEE from sparrows to cowbirds nine, twelve, and sixteen days after taking infective meals. The species feeds readily on birds and on most large vertebrates in its range.

Evidence incriminating *A. vexans* as a natural vector of virus:WEE is mounting (Table 36) for woodland foci of northern USA-48 and southern Canada. Shemanchuk (1969) obtained the agent from this mosquito in Alberta. In Saskatchewan McLintock and Burton (1967) incriminated this mosquito in avian-culicine orbits in interepidemic years. Involved were barn swallows, shrike, magpie, and Franklin's gull, together with *A. vexans* among other mosquitoes. Burroughs and

TABLE 36

Arboviruses Associated with *A. vexans*

Virus	Infection Lab	Infection Wild	Transmission Lab	Transmission Wild	Location	Citation
Alphavirus sp.	. .	+	. .	. .	Kazakhstan	Ananyan, 1964
Bunyamwera sp.	. .	+	. .	. .	Wisconsin	Anslow et al., 1969
California encephalitis	. .	+	. .	. .	Brit. Columbia	McLean, 1970
Tahyna	+	. .	+	. .	Russia	Bardos & Jakubik, 1961
Tahyna	. .	+	. .	. .	Czechoslovakia	Bardos & Danielova, 1959; Malkova et al., 1965
Tahyna	+	. .	+	. .	Europe	Simkova et al., 1960
Tahyna	. .	. .	+	. .	Austria	Aspöck & Kunz, 1967
Tahyna	+	. .	. .	. .	Europe	Danielova, 1966
Getah[a]	. .	+	. .	. .	Japan	Shichijo et al., 1970
Getah[a]	. .	+	. .	. .	Okinawa	Hurlbut & Nibley, 1964
Getah[a]	. .	+	. .	. .	Japan	Scherer et al., 1962
Japanese B encephalitis[a]	. .	+	. .	. .	Japan	Shichijo et al., 1968
Japanese B encephalitis[a]	+	. .	+	. .	. .	Hodes, 1946
Eastern equine encephalitis	+	. .	+	. .	USA	Davis, 1940
Eastern equine encephalitis	+	+	+	. .	Connecticut	Wallis et al., 1960
Eastern equine encephalitis	. .	+	. .	. .	Alabama	Sudia et al., 1968
Eastern equine encephalitis	+	. .	+	. .	USA	Chamberlain, 1958
Jamestown Canyon	. .	+	. .	. .	California	Sudia et al., 1971
LaCrosse	. .	+	. .	. .	USA-48	Sudia et al., 1971
?ittatus	. .	+	. .	. .	Iowa	Wong et al., 1971
?stern equine encephalitis	. .	+	. .	. .	Alberta	Shemanchuk, 1969
Western equine encephalitis	. .	+	. .	. .	Minnesota	Burroughs & Burroughs, 1954
Western equine encephalitis	. .	. .	+	. .	USA	Chamberlain, 1958
Western equine encephalitis	. .	+	. .	. .	Minnesota	Olson et al., 1961
Western equine encephalitis	. .	+	. .	+	Saskatchewan	McLintock & Burton, 1967
West Nile	. .	+	. .	. .	USSR	Berezin et al., 1971
Omsk hemorrhagic fever	. .	+	. .	. .	USSR	Casals et al., 1970

[a] *A. vexans nipponii.*

Burroughs (1954) isolated virus:WEE from wild *A. vexans* from Wright County, Minnesota. This mosquito becomes infected readily, and it transmits the pathogen readily (Chamberlain, 1958).

Evidence is mounting (Table 36) that *A. v. nipponii* is significant in the feral reservoir of Japanese B encephalitis in Okinawa and other parts of Japan (Shichijo *et al.*, 1968, 1970; Scherer *et al.*, 1962; Iha, 1971). One strain has been isolated from *nipponii* near Nagasaki (Hayashi *et al.*, 1970).

Other viruses are increasingly associated with *A. vexans* (Table 36). Among those now known to be associated in the wild are Getah in Okinawa and elsewhere in Japan, Jamestown Canyon in California, LaCrosse and trivittatus in central USA-48. An undetermined Bunyamwera virus was obtained in Wisconsin, and an Alphavirus was obtained in Kazakhstan.

The filariid of domestic importance most likely to be significantly associated with *A. vexans* is *Dirofilaria immitis*. Bemrick and Sandholm (1966) found naturally infected ones in Minnesota and indicated that the species is a field vector. Summers (1943) infected ten of twenty-six *A. vexans* but found development only to the sausage stage. According to Yen (1938), this is a fairly good candidate for vector status. Infection rates were high, and 25 of 129 infected ones survived to produce mature larvae. Rate of infection of one mosquito ranged from 2 to 76 larvae. One specimen dissected on day 12 had infective larvae in the head. Laboratory infections of four to thirty larvae were found in 80 percent of fed mosquitoes, and mature larvae appeared in the labium (Hu, 1931).

Newton *et al.* (1945) found that *Wuchereria bancrofti* could infect this species in the laboratory. It could penetrate the gut wall and lodge in the thoracic muscles. None developed to an infective stage under the conditions imposed.

Toxinosis. Regulation of this species came into prominence with the availability of the persistent toxicants from DDT to methoxychlor. All have proved toxic in field and laboratory, according to numerous reports from abatement districts. For the most part the toxicants have been dispensed from thermal and atomizing devices. In closed spaces such as rooms they have been effective, as are volatile materials that are variously released. Diurnal applications made at pre-dispersal aggregation sites have been notably effective, particularly in the spring. They are toxic enough at all times outdoors, but no delivery system assures contact between adult mosquitoes and the toxicant after dispersal has occurred. Toxinosis has never been the deficiency as a regulator outdoors; timing and inadequate delivery have been.

Mitchell *et al.* (1970) noted that aerial "ultra-low volume" applications of malathion in western Texas caused significant population reductions in collections made during days 1-3 or 4, but not day 5 or beyond after treatment. In summary, the authors conclude that the relief afforded was not commensurate with the effort.

Many of the toxic products (notably those containing DDT and methoxychlor), when deposited on vegetation in post-dispersal aggregation sites, will cause *A. vexans* to depart and may prevent their return for one to four days. The principle has been successfully exploited by mosquito abatement districts to deny aggregation near dwellings, camp sites, parks, and picnicking sites. Some odorous fractions of petroleum, when free to volatilize in the open air, will act to drive this mosquito from yards and similar situations. Smoke derived from petroleum products will do the same.

Resistance. No evidence of significant resistance to toxicants, especially persistent ones such as the chlorinated hydrocarbons, is known (Sutherland and Hagmann, 1963, and others).

Mortality factors. No reports on the relative importance of mortality factors in the environment have been seen. The cumulative effects of natural causes decimate a brood of adults in the field in two to five weeks. Little loss occurs between emergence and at least one oviposition effort, especially where host animals and oviposition sites are proximal. Apparently survival for extensive initial dispersal flights is good, but it is reasonable to postulate that as the distance between emergence sites, feeding sites, and oviposition sites is lengthened, progressive decimation ensues. Abundance and availability of hosts seem to be ample for maximum survival. Drought may decrease availability of oviposition sites long enough to cause many adults to die before ovipositing. A combination of drought, excessive diurnal temperature in perching sites, and wind increase mortality sufficient to affect reproductivity. Toxic applications to the imaginal stage have shown no significant effect on ability of a brood to regenerate itself.

Colonization

Colonization of *Aedes vexans* is possible with a population of any stage brought from the field. Two phases of the life history require special attention. Adults must be manually copulated for consistent maintenance, and larvae must be fed at a rate that provides adequate nutrition and in a manner to avoid contamination of the water by scum-forming microbes. All floodwater mosquitoes are extremely sensitive to toxic effects of scum-forming biota. Cages for handling are simple, compact, and permit rearing of large numbers if necessary.

Oviposition. Eggs may be obtained easily by allowing caged females to oviposit on strips of cheesecloth lying on wet cellucotton. Cages that permit convenient transfer and feeding of adults may be made of lucite and netting (size 15 x 2.5 x 2.5 cm). The netting should be nylon tulle with mesh large enough to permit females to deposit eggs outside the cage. Walls and ends are 3 mm thick, and top and bottom are of nylon tulle. Ten or fifteen mosquitoes are introduced through a hole in one end; the hole is closed by stopper. The cage is placed in a pan containing six layers of cheesecloth on a pad of wet cellucotton 1.5 cm thick. The pan is slanted; water is kept at a level that insures a gradient of moisture in the cellucotton from that of a water-logged state to a moist one. Room temperature of 25° and pad temperature of 22-25° are suitable. Continuous lighting is provided. Mosquitoes are kept in small cages that may be moved from oviposition or feeding site to facilitate handling.

Mosquitoes are allowed to feed daily on human blood (alternative hosts: rabbit, rodent, or avian blood) by placing the cage on any part of the body that is convenient. Honey diluted 1:10 with water provides carbohydrate nutrition when made available in a ball of cellucotton.

Eggs are deposited outside the cage on the cheesecloth mat lying on the cellucotton pad. They are tucked into the mesh wherever moisture level is attractive. A mat of six single layers of cheesecloth is ample to insure that all eggs are in the cloth and not in the cellucotton pad.

Storage of eggs. Eggs are removed daily from the mats by unfolding the cloth in water 10+ cm deep. Water containing eggs is passed through a sieve (100 meshes to the inch) to concentrate the eggs. They are then washed into a small casserole under water. A dropping pipet is then used to transfer eggs to filter-paper slips on wet cellucotton pads in petri dishes. Eggs are held at 25° for at least ten days for embryonation. After about ten days the papers bearing the eggs are put on moist (not glistening) pads in petri dishes for holding at 25° or 4°, depending on intended use. The latter is the usual storage temperature. Eggs may be held for months in a moist chamber at 4°.

Hatching. The final act of hatching is stimulated by lowering the level of dissolved oxygen. As a routine oxygen is reduced biotically by growth of microbes in a tube. Powdered nutrient broth is dissolved in aerated water at a dilution of 1:1,000 for hatching at 25°, at 1:2,000 for 30°, and at 1:500 for 21°. Up to 100 eggs are put in a glass vial 3 mm inside diameter, and the vial is submerged in the larger vial of hatching media. At 25° conditioned eggs hatch in one to three hours.

Eggs brought in from the field or those subjected for long intervals to 4° hatch erratically unless held at 25° for ten to fourteen days. The time may be shortened to four or five days by exposure to 30°. This latter temperature is suggested for obtaining uniformly high hatching rate for eggs from natural sites (see *Egg: Hatching*).

Rearing. Larvae are reared in shallow water in covered pans. The standard temperature is 25°. Food of choice is a slurry of pulverized "fish food" (Tetramin is the commercial product used locally) in the manner proposed by Brust (1971). Deionized water with 10-20 ml of oak-leaf infusion is placed in the rearing pans to a depth of 1 cm. Eight pinches of washed sand are dispersed in the pans. Enough slurry is dropped with a pipet on the spots of sand to nourish the larvae for a day, although a slight excess is not objectionable. Larvae one or two hours old are concentrated in a 100-mesh stainless steel basket 25 mm in diameter with sand and food near one end of the pan. They are thus kept for the first day to assure adequate exposure to food. After day 1 larvae are transferred by pipet to free range in the pan. Food is added daily, and no change of water is required. When larvae are in instar 3, more deionized water with oak-leaf infusion may be added to a depth of 2-3 cm.

Pupae are transferred by pipet to clean, deionized water in emergence cages (size: 18 x 18 x 18 cm). These cages are made of lucite with a door in the side and a feeding slot in the top. A cellucotton wick in a well of deionized water provides humidity. A pledget of cotton saturated with honey-water dilution 1:10 placed in the feeding slot provides carbohydrate for adults.

Mating. Colony maintenance depends on manual copulation of adults. The method currently in use is a modification of the first one developed in this laboratory (McDaniel and Horsfall, 1957). Adults are held seven to ten days in emergence cages in a lighted room. For the first two to five days, the holding temperature is 25°, and subsequently the temperature is 21°. Mating is more dependable if the holding temperature is 21° for a day before mating.

A male is prepared for mating by placing it in a vial, then inactivating it momentarily in carbon dioxide, and finally decapitating it and mounting it on a glass microscope slide. The mesonotum is impaled in a small spot of a casein glue so that the ventral side is uppermost and the abdomen is slightly elevated and free. Three or four males may be glued to one slide. The venter of the rotated gonapophyses must be fully exposed and not twisted to one side.

Females are wholly inactivated by carbon dioxide (2-5 minutes). A tube 2 cm in diameter fitted with a screen to make a bottom pervious to carbon dioxide is the anesthetizing chamber, which is connected by a tube to a tank of compressed carbon dioxide. Two to five females may be held at one time. Deactivated females are transferred to a surface from which they are picked up with a suction pipet attached to a small vacuum pump of the sort used in industry to handle small parts during assembly of machinery. The pipet is made by shortening a hypodermic needle and bending it so that the tip can be placed to the mesonotum behind the head. The immobilized female is held venter upward and at an angle of about 120° with that of the male. When in position, copulation is accomplished in ten to thirty seconds.

Several factors which might influence the success of induced copulation have been reported by Horsfall and Taylor (1967), Jones and Wheeler (1965), and Spielman (1964). Other than the requisite for vigorous mosquito stock, influencing factors include consideration of the age of males selected for mating, pre-mating holding temperatures, the sequence of pairing virgin females with single males, and the selection of an anesthetic to immobilize females during the induced copulation period. Horsfall and Taylor (1967) found that exposure of *A. vexans* to 16° diminished the mating response of males and that a high order of insemination could be obtained by use of males fourteen days old which were held at 23° prior to mating. Jones and Wheeler (1965) observed that males of *A. aegypti*, when subjected to rapid successive copulations, ejaculated progressively fewer spermatozoa with each successive female and that copulation could occur without insemination. A variety of anesthetics have been used by different investigators. Jones and Wheeler (1965b) demonstrated that spermathecal filling can occur in females of *A. aegypti* which are externally immobilized immediately following manual copulation by exposure to nitrogen, carbon dioxide, ether, and cyclopropane. Spielman (1964), working with a colony of *A. aegypti* from Johns Hopkins University, noted that carbon dioxide appeared to interfere with manual copulation. He adopted nitrogen as the anesthetic of choice for males and ether for females. In addition, he observed that females anesthetized with ether had an extended and very irregular copulatory period, but no information was provided on rates of insemination for this group. Horsfall and Taylor (1967) found chloroform to be a satisfactory anesthetic for immobilizing females, but they cautioned against overexposure and observed that females of *A. vexans* anesthetized with chloroform seldom feed on blood during the first six to twelve hours after recovery.

During this investigation the influence of anesthetics on rates of insemination was investigated by anesthetizing virgin females with nitrogen, carbon dioxide, and chloroform prior to inducing copulation. Single decapitated males were mated with virgin females for three repetitive copulations; rates of insemination were determined by dissection and examination of spermathecae for presence of spermatozoa. The efficiency of insemination varied according to the anesthetic used prior to copulation and by the sequence of mating virgin females with single males. Rates of insemination were consistently high for females anesthetized with nitrogen (87%), intermediate for females anesthetized with carbon dioxide (70%), and low for females anesthetized with chloroform (57%).

Results obtained during this investigation would indicate that the efficiency of insemination of virgin females of *A. vexans* during induced copulation is enhanced by use of an anesthetic which allows rapid recovery of adults. Females anesthetized with nitrogen recover during coitus and require special care to prevent them from escaping after mating. On presentation of females immobilized with nitrogen to decapitated males, copulation must be induced within five to eight seconds, or females will contract their genital segments, thereby making copulation impossible. Females anesthetized with carbon dioxide recover in approximately twenty seconds, and there is sufficient time to present a female to a second male if copulation cannot be induced initially. During this study females immobilized with carbon dioxide normally did not recover during coitus. With chloroform females require from two to four minutes to recover, and they remain relatively inactive for an additional ten minutes.

Summary

Aedes vexans is a major pest of man and livestock, particularly on the prairies and plains of Holarctica. It is a prolific species that arises from transient pools, flooded river valleys, and margins of fluctuating lakes, swamps, and sloughs. It may be a major pest inland from lowland coastal areas where streams and runoff water flood grasslands on the land side of the tidal zone. *A. vexans vexans* coincides with *Psorophora* spp. over a wide range in North America, with coastal aedines to a degree in eastern USA, and with melt-water aedines to the north. *A. vexans nocturnus* occurs in Pacific Oceania from Hawaii to Samoa, and *A. vexans nipponii* is a pest in Japan and eastern Asia. As vectors of pathogens, *A. vexans* is significant in zoonotic transmission of

Tahyna and encephalitic viruses. It may be important as a juvenile host for *Dirofilaria immitis* as well.

This is a multivoltine species in the usual sense. Repetitive generations occur without a mandatory hiatus characteristic of many aedines. It may also act as a univoltine species, and eggs enter a state where hatching is reduced or prevented when adults are held under light conditions simulating short days of the fall prior to oviposition. Frequency of generations or broods is related more to frequency of flooding of oviposition sites than to any other cause.

Aedes vexans is a floodwater mosquito because eggs are deposited on soil and remain until inundated by water at temperatures of about 15° or above. Eggs hatch shortly after being submerged, and larvae mature before biota of permanent water become established. Water fouled by sewage and certain industrial wastes is inimical to larvae. Rearing media in the laboratory require that water be devoid of scum-forming microbes.

Density of eggs at oviposition sites is a function of durability of a level of moisture attractive to gravid females. The most favorable sites are sandy soils that become firm at the surface and are kept moist by frequent light rain or by a layer of loose herbal detritus that dries slowly. In woodland sites soils attractive to ovipositing females are provided over a wide zone below that of prior maximum flood. Width of the zone of oviposition may be variably wide and the population variably dense according to the rate of decline in water level following a period of flooding.

Hatching of eggs takes place during the interval of rising water level in the field. Eggs are first submerged; then the oxygen declines at the surface of the soil, and larvae rupture the shell by means of hatching spines. The final hatching stimulus for an embryo to rupture its eggshell comes from decrease in oxygen. Once the shell is open, water floods the embryo, and it simply expands to free itself from the shell.

Embryos survived in eggs at room temperature (25°) for nearly a year. When held at 4° in a moist chamber, survival for two years has occurred. In the field survival has been variously stated to extend for several years. Eggs stored in the field under controlled conditions have survived until August of the second summer, but hazards such as burial, predation, or hatching during prolonged periods of rain prevent significant survival beyond the second summer.

Larvae change position in their developmental sites according to activity and changes in movement of the water. When in instar 1 they tend to be marginal, especially in water devoid of emergent herbs and grass. Older larvae disperse as water recedes, especially on flood plains of rivers. Older larvae and pupae may move several kilometers when in the channel of streams and rivers.

Since larvae appear in a wide variety of sites during the ice-free part of the year over most of Holarctica, *A. vexans* may be present in conjunction with a vast array of species of aedines. It may be coincident with many species in permanent water of lakes and in impoundments subject to fluctuating levels.

Adults may be present during all ice-free months but are erratically abundant according to extent of sources, distance from sources, availability of shelter, and availability of food. Suburbs are particularly invaded because of attraction to light and availability of food and shelter. Adults fly from developmental sites for distances of a few to hundreds of kilometers, according to vagaries of meteorological conditions. For the usual appetential movement they fly near the ground, but flights may occur up to a thousand or more meters above the ground. Exodus flights begin particularly at dusk and may take place during one or more nights, according to the time of emergence and according to the time spread in emergence created by variations in submergence of larval sites. Exodus flights may be thwarted when populations emerge in the spring before the occurrence of early evening temperatures conducive to flight.

Colonies of *Aedes vexans* may be maintained by rearing larvae in water 20 mm deep and with a daily diet of "fish food." Adults must be manually copulated. This is easily done so that populations of any level may be easily maintained.

Hazards to survival of this species in nature are largely imposed on the egg stage and to some extent on the larval stage. From the time they are deposited, burial by burrowing and trampling animals and by siltation during winter floods may keep up to 75 percent of eggs from hatching. Sites subject only to premature flooding during winter or spring thaw do not permit eggs to hatch. A season of so little rainfall that no flooding occurs results in no hatching. Two adversities may be experienced by larvae: water may remain too short a time, or flooding may occur too late in the fall to permit development.

FIGURES

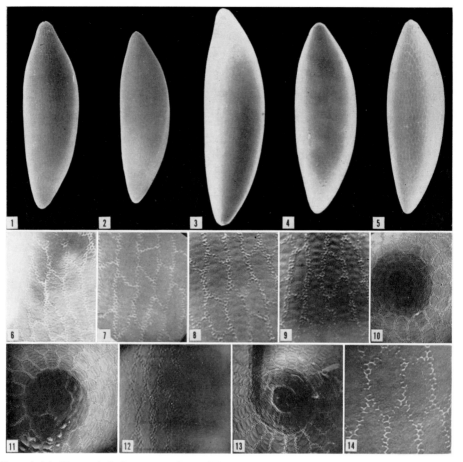

FIGS. 1-14: Eggs of *Aedes vexans* from different geographical areas (from Horsfall *et al.*, 1970).

Figs. 1-5: Whole eggs of *Aedes vexans*. Dorsal side to left, anterior end uppermost. 90×
Figs. 6-9, 12, 14: Chorionic reticulation. Figs. 6-9, 12, 450×; Fig. 14, 1080×
Figs. 10, 11, 13: Micropylar sculpturing on anterior end. 450×

Fig. 1. Mermet, Ill.

Fig. 8. Allerton Park, Ill. Long form.

Fig. 2. Olney, Ill.

Fig. 9. Allerton Park, Ill. Short form.

Fig. 3. Allerton Park, Ill. Long form.

Fig. 10. Flin Flon, Manitoba.

Fig. 4. Allerton Park, Ill. Short form.

Fig. 11. Florence, Ala.

Fig. 5. Flin Flon, Manitoba.

Fig. 12. Florence, Ala.

Fig. 6. Mermet, Ill.

Fig. 13. St. Joseph, Ill.

Fig. 7. Olney, Ill.

Fig. 14. St. Joseph, Ill.

F<small>IG</small>. 15. Removing soil bearing eggs of floodwater mosquitoes from a woodland site in late fall.

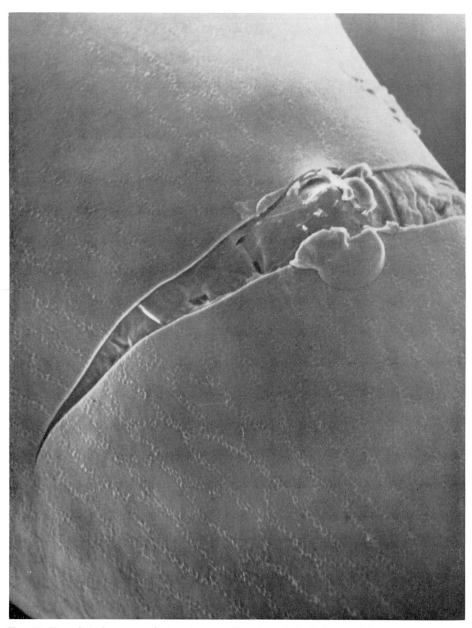

FIG. 16. Egg of *Aedes vexans* hatching (from Horsfall *et al.*, 1969). 899×

Fig. 17. Larva of instar 4; lateral view. 20×

108

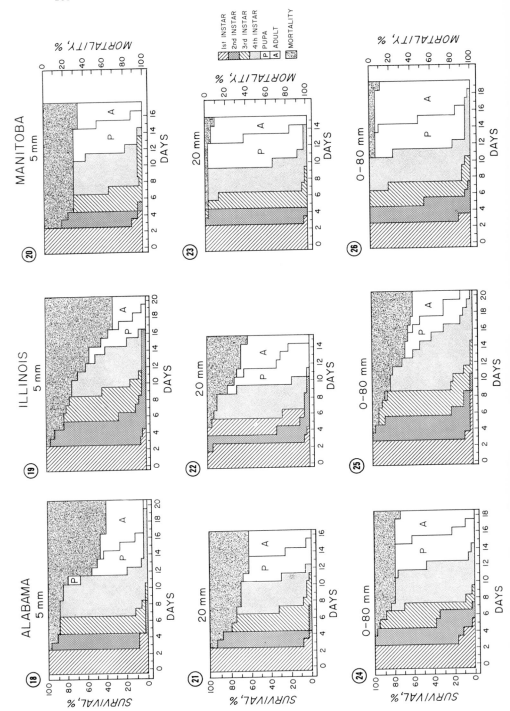

FIGS. 18-26. Bionomic charts showing development and mortality of cultures of *Aedes vexans* obtained from different geographical areas when reared at varying depths of water at constant 18° and constant light and fed live yeast.

Figs. 18-20: Water depth: 5 mm
    Fig. 18. Alabama strain.
    Fig. 19. Illinois strain.
    Fig. 20. Manitoba strain.

Figs. 21-23: Water depth: 20 mm
    Fig. 21. Alabama strain.
    Fig. 22. Illinois strain.
    Fig. 23. Manitoba strain.
Figs. 24-26: Water depth: 0-80 mm
    Fig. 24. Alabama strain.
    Fig. 25. Illinois strain.
    Fig. 26. Manitoba strain.

Figs. 27–32: Bionomic charts showing development and mortality of cultures of *Aedes vexans* obtained from Illinois when reared at varying depths of water at constant 25° and 30° and constant light and fed live yeast.

Figs. 27–29: Culture of *Aedes vexans* at 25°.
    Fig. 27. Water depth: 5 mm
    Fig. 28. Water depth: 20 mm
    Fig. 29. Water depth: 0 to 80 mm

Figs. 30–32: Culture of *Aedes vexans* at 30°.
    Fig. 30. Water depth: 5 mm
    Fig. 31. Water depth: 20 mm
    Fig. 32. Water depth: 0 to 80 mm

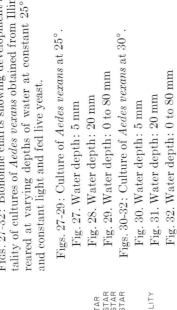

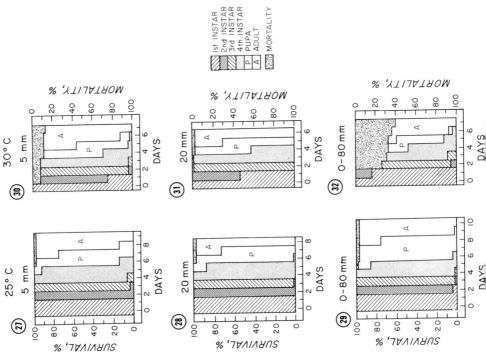

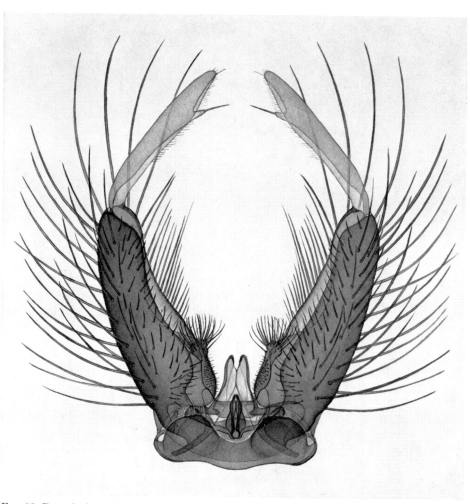

F_{IG}. 33. Dorsal view of caudal appendages of male of *Aedes vexans*. 80×

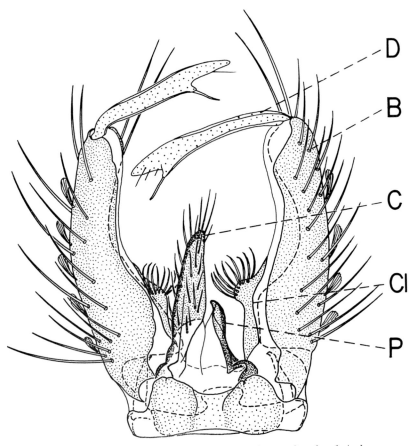

FIG. 34. Dorsal view of anomalous caudal appendages of male of *Aedes vexans* 80× (from Cupp and Fowler, 1969).

D: dististyle (♂) (Gonostylus)

B: basistyle (♂) (Gonocoxite)

C: cercus (♀)

Cl: claspette (♂)

P: paraproct (♂)

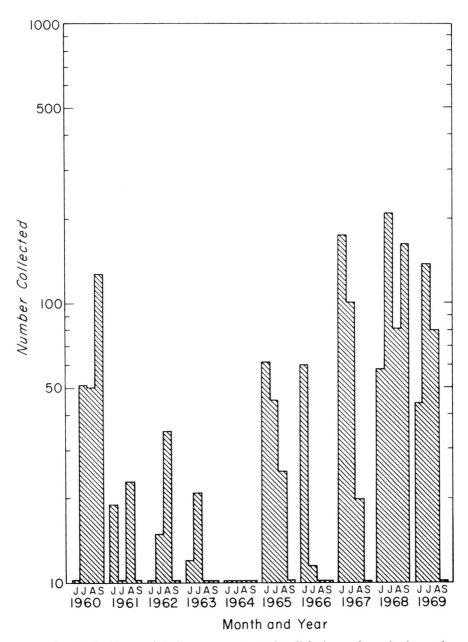

FIG. 35. Incidence of *Aedes vexans* attracted to light in a sylvan site in northern suburbia of metropolitan Cook County, Illinois. Each bar represents the highest nightly collection for each month of summer 1960-69. (Courtesy North Shore Mosquito Abatement District.)

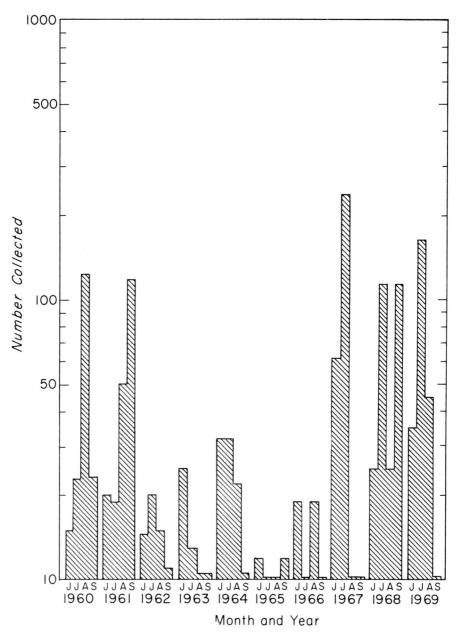

F<small>IG</small>. 36. Incidence of *Aedes vexans* attracted to light in a sylvan site in western suburbia of metropolitan Cook County, Illinois. Each bar represents the highest nightly collection for each month of summer 1960-69. (Courtesy Des Plaines Valley Mosquito Abatement District.)

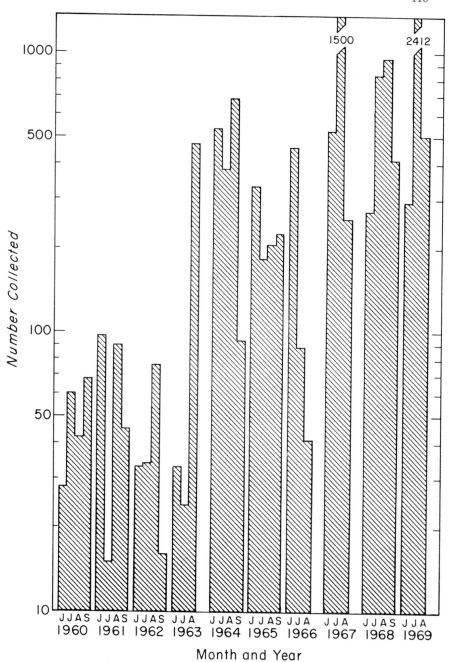

F<small>IG</small>. 37. Incidence of *Aedes vexans* attracted to light in a sylvan site in southern suburbia of metropolitan Cook County, Illinois. Each bar represents the highest nightly collection for each month of summer 1960-69. (Courtesy South Cook County Mosquito Abatement District.)

Fig. 39. Light trap based on design of the New Jersey trap with devices for killing by heat and for separating specimens according to time of collection. (For specifications and design, see Horsfall, 1962.)

Fig. 38. Stationary interceptor unit of four traps. (For specifications, see Horsfall, 1961.)

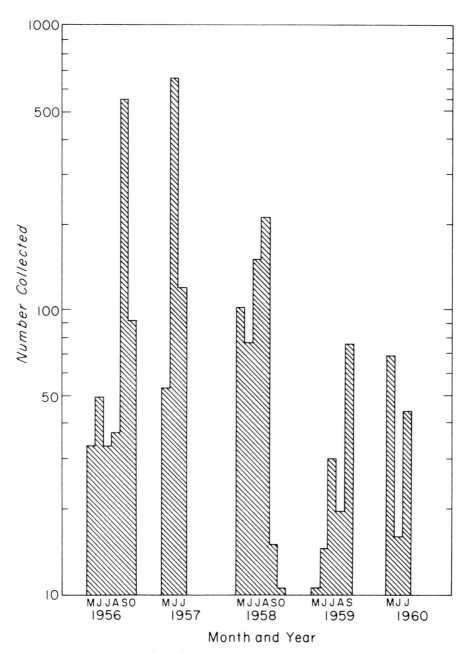

Fig. 40. Incidence of *Aedes vexans* attracted to light in a sylvan site north of Urbana, Illinois. Each bar represents highest nightly collection for each month of summer 1956-60.

REFERENCES CITED

Ackerman, R., W. Spithaler, W. Profittlich, and D. Spieckerman. 1970. Über die Verbreitung von Viren der California-Encephalitis-Gruppe in der Bundesrepublik Deutschland. *Deutsche medizinische Wochenschrift* 95:1507-1513.

Adams, C. F., and W. M. Gordon. 1943. Notes on the mosquitoes of Missouri (Diptera: Culicidae). *Entomological News* 54:232-235.

Allen, J. R., and A. S. West. 1962. Collection of oral secretions from mosquitoes. *Mosquito News* 22:157-159.

Ananyan, S. A. 1964. Preliminary serological and virological data on the isolation of arbo-A viruses in eastern parts of the U.S.S.R. *Acta Virologica*, Prague, 8:93.

Anderson, J. F. 1968. Influence of photoperiod and temperature on the induction of diapause in *Aedes atropalpus* (Diptera: Culicidae). *Entomologia Experimentalis et Applicata* 11:321-330.

Annand, P. N. 1941. Report of the chief. *Report of the U.S. Bureau of Entomology and Plant Quarantine* 1940-41:91.

Anslow, R. O., W. H. Thompson, P. H. Thompson, G. R. De Foliart, O. Papadopoulos, and R. P. Hanson. 1969. Isolation of Bunyamwera-group viruses from Wisconsin mosquitoes. *American Journal of Tropical Medical Hygiene* 18:599-608.

Aspöck, H., and C. Kunz. 1967. Untersuchungen über die Ökologie des Tahyna-Virus. *Zentralblatt für Bakteriologie*, Abt. I Orig. 203:1-24.

Baker, M. 1961. The altitudinal distribution of mosquito larvae in the Colo-

119

rado Front Range. *Transactions of the American Entomological Society* 87:231-246.

Bardos, V., and V. Danielova. 1959. The Tahyna virus — a virus isolated from mosquitoes in Czechoslovakia. *Journal of Hygiene, Epidemiology, Microbiology, and Immunology* 3:264-276.

Bardos, V., and J. Jakubik. 1961. Experimental infection of pigs with Tahyna virus. *Acta Virologica*, Prague, 5:228-231.

Barnes, R. C., H. L. Fellton, and C. A. Wilson. 1950. An annotated list of the mosquitoes of New York. *Mosquito News* 10:69-84.

Barr, A. R. 1954. A note on the chaetotaxy of *Aedes vexans* (Meigen, 1830). *Mosquito News* 14:24-25.

―――. 1958. The mosquitoes of Minnesota (Diptera: Culicidae: Culicinae). *Technical Bulletin of the Minnesota Agricultural Experiment Station* 228. 154 pp.

Bates, M. 1949. *The natural history of mosquitoes.* New York: The Macmillan Co. xv + 379 pp.

Beadle, L. D. 1963. The mosquitoes of Isle Royale, Michigan. *Proceedings, New Jersey Mosquito Extermination Association* 50:133-139.

Bemrick, W. J., and H. A. Sandholm. 1966. *Aedes vexans* and other potential mosquito vectors of *Dirofilaria immitis* in Minnesota. *Journal of Parasitology* 52:762-767.

Benson, R. L. 1936. Diagnosis and treatment of sensitization to mosquitoes. *Journal of Allergy* 8:47-57.

Bidlingmayer, W. L., and H. F. Schoof. 1957. The dispersal characteristics of the salt-marsh mosquito, *Aedes taeniorhynchus* (Wiedemann), near Savannah, Georgia. *Mosquito News* 17:202-212.

Bishopp, F. C., E. N. Cory, and A. Stone. 1933. Preliminary results of a mosquito survey in the Chesapeake Bay section. *Proceedings of the Entomological Society of Washington* 35:1-6.

Blickle, R. L. 1952. Notes on the mosquitoes (Culicinae) of New Hampshire. *Proceedings, New Jersey Mosquito Extermination Association* 39:198-202.

Boddy, D. W. 1948. An annotated list of the Culicidae of Washington. *Pan-Pacific Entomologist* 24:85-94.

Bodman, M. T., and N. Gannon. 1950. Some habitats of eggs of *Aedes vexans*. *Journal of Economic Entomology* 43:547.

Bohart, R. M. 1954. Identification of first stage larvae of California *Aedes* (Diptera, Culicidae). *Annals of the Entomological Society of America* 47:355-366.

Borg, A., and W. R. Horsfall. 1953. Eggs of floodwater mosquitoes. II. Hatching stimulus. *Annals of the Entomological Society of America* 46:472-478.

Breeland, S. G., A. W. Buzicky, E. Pickard, and W. I. Barton. 1965. Comparative observations on winter survival and hatching of *Aedes vexans* eggs in two localities — Florence, Alabama, and St. Paul, Minnesota. *Mosquito News* 25:374-384.

Breeland, S. G., and E. Pickard. 1964. Insectary studies on longevity, blood-feeding, and oviposition behavior of four floodwater mosquito species in the Tennessee Valley. *Mosquito News* 24:186-192.

————. 1967. Field observations on twenty-eight broods of floodwater mosquitoes resulting from controlled floodings of a natural habitat in the Tennessee Valley. *Mosquito News* 27:343-358.

Bresslau, E. 1917. Beiträge zur Kenntnis der Lebensweise unserer Stechmücken über die Eiablage der Schnaken. *Biologisches Zentralblatt* 37:507-531.

Breyev, K. A. 1963. The effect of various light sources on the numbers and species of blood-sucking mosquitoes (Diptera: Culicidae) collected in light-traps. *Entomological Review, Washington* 42:155-168.

Brust, R. A. 1968. Effect of starvation on molting and growth in *Aedes aegypti* and *A. vexans*. *Journal of Economic Entomology* 61:1570-1572.

————. 1971. Laboratory mating in *Aedes dianteus* and *Aedes communis* (Diptera: Culicidae). *Annals of the Entomological Society of America* 64:234-237.

————, and R. A. Costello. 1969. Mosquitoes of Manitoba II. The effect of storage temperature and relative humidity on hatching of eggs of *Aedes vexans* and *Aedes abserratus* (Diptera: Culicidae). *Canadian Entomologist* 101:1285-1291.

Burgess, L., and W. O. Haufe. 1960. Stratification of some prairie and forest mosquitoes in the lower air. *Mosquito News* 20:341-346.

Burroughs, A. L., and R. N. Burroughs. 1954. A study of the ecology of western equine encephalomyelitis virus in the upper Mississippi River Valley. *America Journal of Hygiene* 60:27-36.

Callot, J., and D. Van-Ty. 1944. Contribution a l'étude des moustiques Francais Culicides de Richelieu (Indre-et-Loire). *Annales de Parasitologie Humaine et Comparée* 20:43-66 + 2 pls.

Carpenter, S. J., and W. J. LaCasse. 1955. *Mosquitoes of North America North of Mexico*. University of California Press, Berkeley. vi + 360 pp. + 127 pls.

Casals, J., B. E. Henderson, H. Hoogstraal, K. M. Johnson, and A. Shelokov. 1970. A review of Soviet viral hemorrhagic fevers, 1969. *Journal of Infectious Diseases* 122:437-453.

Chamberlain, R. W. 1958. Vector relationships of the arthropod-borne encephalitides in North America. *Annals of the New York Academy of Science* 70:312-319.

Chang, T. L. 1939. Mosquitoes of Hunan Province with special reference to *Anopheles*. *China Medical Journal* 56:52-62.

Chapman, H. C., T. B. Clark, D. B. Woodard, and W. R. Kellen. 1966. Additional mosquito hosts of the mosquito iridescent virus. *Journal of Invertebrate Pathology* 8:545-546.

Chapman, H. C., D. B. Woodard, and J. J. Petersen. 1967. Pathogens and parasites in Louisiana Culicidae and Chaoboridae. *Proceedings, New Jersey Mosquito Extermination Association* 54:54-60.

Chow, C. Y. 1949. Culicine mosquitoes collected in western Yunnan, China during 1940-1942 (Diptera, Culicidae). *Proceedings of the Entomological Society of Washington* 51:127-132.

Christensen, G. R., and F. C. Harmston. 1944. A preliminary list of mosquitoes of Indiana. *Journal of Economic Entomology* 37:110-111.

Christophers, S. R. 1960. *Aedes aegypti* (L.) — *the yellow fever mosquito* —

its life history, bionomics and structure. Cambridge: University Press. 739 pp.

Clarke, J. L. 1937. New and significant experiences in mosquito control in the Desplaines Valley mosquito abatement district. *Proceedings, New Jersey Mosquito Extermination Association* 24:112-126 + 1 pl.

————. 1943. Studies on the flight range of mosquitoes. *Journal of Economic Entomology* 36:121-122.

Clarke, J. L., Jr., and F. C. Wray. 1967. Predicting influxes of *Aedes vexans* into urban areas. *Mosquito News* 27:156-163.

Clements, A. N. 1963. *The physiology of mosquitoes.* Pergamon Press, New York. ix + 393 pp.

Costello, R. A., and R. A. Brust. 1969. A quantitative study of uptake and loss of water by eggs of *Aedes vexans* (Diptera: Culicidae). *Canadian Entomologist* 101:1266-1269.

————. 1971. Longevity of *Aedes vexans* under different temperatures and relative humidities in the laboratory. *Journal of Economic Entomology* 64: 324-325.

Couch, J. N., and H. R. Dodge. 1947. Further observations on *Coelomomyces* parasitic on mosquito larvae. *Journal of the Elisha Mitchell Scientific Society* 63:69-79 + 6 pls.

Craig, G. B., Jr. 1955. Preparation of the chorion of eggs of aedine mosquitoes for microscopy. *Mosquito News* 15:228-231.

————. 1956. Classification of eggs of nearctic aedine mosquitoes (Diptera: Culicidae). Ph.D. thesis, University of Illinois, Urbana, Ill. 128 pp.

Cupp, E. W., and H. W. Fowler, Jr. 1969. Anomalous dimorphism in *Aedes vexans. Annals of the Entomological Society of America* 62:914-917.

Curry, D. P. 1939. A documented record of a long flight of *Aedes sollicitans. Proceedings, New Jersey Mosquito Extermination Association* 29:36-39.

Danielova, V. 1962. Multiplication dynamics of Tahyna virus in different body parts of *Aedes vexans* mosquitoes. *Acta Virologica,* Prague, 6:227-230.

————. 1966. Quantitative relationships of Tahyna virus and the mosquito *Aedes vexans. Acta Virologica,* Prague, 10:62-65.

Darsie, R. F., D. MacCreary, and L. A. Stearns. 1951. An annotated list of the mosquitoes of Delaware. *Proceedings, New Jersey Mosquito Extermination Association* 38:137-146.

Dashkina, N. G., and D. B. Tsarichkova. 1965. Copulation of some mosquito species of the genus *Aedes* in the laboratory [in Russian]. *Meditsinskaya parazitologiya i parazitarnÿe bolezni Moskva* 34:235-[Trans. Can. NRC# C-8268].

Davis, W. A. 1940. A study of birds and mosquitoes as hosts for the virus of eastern equine encephalomyelitis. *American Journal of Hygiene* 32:(sec. C) 45-59.

Dickinson, W. E. 1944. Mosquitoes of Wisconsin. *Bulletin of the Public Museum, Milwaukee* 8:269-365.

Dodge, H. R. 1966. Studies on mosquito larvae II. The first-stage larvae of North American Culicidae and the world Anophelinae. *Canadian Entomologist* 98:337-393.

Dorer, R. E., W. E. Bickley, and H. P. Nicholson. 1944. An annotated list of the mosquitoes of Virginia. *Mosquito News* 4:48-50.

Dorsey, C. K. 1944. Mosquito survey activities at Camp Peary, Virginia. *Annals of the Entomological Society of America* 37:376-387.

Dupree, W. H. 1905. The mosquitoes of Louisiana and their pathogenic possibilities, with remarks upon their extermination. *Medical and Surgical Journal of New Orleans* 58:1-16.

Dyar, H. G. 1902. Notes on mosquitoes in New Hampshire. *Proceedings of the Entomological Society of Washington* 5:140-148.

———. 1917. Notes on the *Aedes* of Montana (Diptera, Culicidae). *Insecutor inscitiae menstruus* 5:104-121.

———. 1921. The mosquitoes of Canada (Diptera, Culicidae). *Transactions of the Royal Canadian Institute* 13:71-120.

———. 1922. The mosquitoes of the United States. *Proceedings of the U.S. National Museum* 62:1-119.

Eckstein, F. 1919. Die Überwinterung unserer Stechmücken. *Biologisches Zentralblatt* 38:530-536.

Edman, J. D., and W. L. Bidlingmayer. 1969. Flight capacity of blood-engorged mosquitoes. *Mosquito News* 29:386-392.

Edman, J. D., and A. E. R. Downe. 1964. Host-blood sources and multiple-feeding habits of mosquitoes in Kansas. *Mosquito News* 24:154-160.

Feemster, R. F., and V. A. Getting. 1941. Distribution of the vectors of equine encephalomyelitis in Massachusetts. *American Journal of Public Health* 31:791-802.

Fellton, H. L., R. C. Barnes, and C. A. Wilson. 1950. New distribution records for the mosquitoes of New England. *Mosquito News* 10:84-91.

Felt, E. P. 1904. Mosquitoes or Culicidae of New York state. *Bulletin of the New York State Museum* 79:241-400.

Ferguson, F. F., and T. E. McNeel. 1954. The mosquitoes of New Mexico. *Mosquito News* 14:30-31.

Filsinger, C. 1941. Distribution of *Aedes vexans* eggs. *Proceedings, New Jersey Mosquito Extermination Association* 28:12-19.

Fowler, H. W., Jr. 1969. Bionomics of *Aedes vexans* (Meigen) (Diptera: Culicidae). Ph.D. thesis. University of Illinois, Urbana, Illinois. 115 pp.

———. 1972. Rates of insemination by induced copulation of *Aedes vexans* (Diptera: Culicidae) treated with three anesthetics. *Annals of the Entomological Society of America* 65:293-296.

Freeborn, S. B., and R. M. Bohart. 1951. The mosquitoes of California. *Bulletin of the California Insect Survey* 1:25-78.

Gerhardt, R. W. 1966. The mosquitoes of South Dakota and their control. *Bulletin of the South Dakota Agricultural Experiment Station* 531:1-80.

Gilot, B. 1969. Répartition altitudinale des moustiques (Dipt.: Culicidae): exemple du Dauphiné du nord (Alpes francaises). *Cahiers de l'Office de la Recherche Scientifique et Technique Outre-Mer, Paris, series Entomologique medicale et Parasitologie* 7:213-233.

Gjullin, C. M. 1937. The female genitalia of the *Aedes* mosquitoes of the Pacific coast states. *Proceedings of the Entomological Society of Washington* 39: 252-266.

————. 1946. A key to the *Aedes* females of America north of Mexico. *Proceedings of the Entomological Society of Washington* 48:215-236.

————, C. P. Hegarty, and W. P. Bollen. 1941. The necessity of a low oxygen concentration for the hatching of *Aedes* mosquito eggs. *Journal of Cellular and Comparative Physiology* 17:193-202.

Gjullin, C. M., W. W. Yates, and H. H. Stage. 1939. The effect of certain chemicals on the hatching of mosquito eggs. *Science* 89:539-540.

————. 1950. Studies on *Aedes vexans* (Meig.) and *Aedes sticticus* (Meig.), flood-water mosquitoes, in the lower Columbia River Valley. *Annals of the Entomological Society of America* 43:262-275.

Glick, P. A. 1939. The distribution of insects, spiders and mites in the air. *Technical Bulletin, U.S. Department of Agriculture* 673:1-150 + 5 pl.

Gunstream, S. E., and R. M. Chew. 1964. Contribution to the ecology of *Aedes vexans* (Diptera: Culicidae) in the Coachella Valley, California. *Annals of the Entomological Society of America* 57:383-387.

Haeger, J. S. 1960. Behavior preceding migration in the salt-marsh mosquito, *Aedes taeniorhynchus* (Wiedemann). *Mosquito News* 20:136-148.

Happold, D. C. D. 1965. Mosquito ecology in central Alberta I. The environment, the species and studies of the larvae. *Canadian Journal of Zoology* 43:794-819.

Harden, F. W., and B. J. Poolson. 1969. Seasonal distribution of mosquitoes of Hancock County, Mississippi, 1964-1968. *Mosquito News* 29:407-414.

Harmston, F. C. 1949. An annotated list of the mosquito records from Colorado. *Great Basin Naturalist* 9:65-75.

————, and D. M. Rees. 1946. Mosquito records from Idaho. *Pan-Pacific Entomologist* 22:148-156.

Harmston, F. C., G. R. Shultz, R. B. Eads, and G. C. Menzies. 1956. Mosquitoes and encephalitis in the irrigated high plains of Texas. *Public Health Reports* 71:759-766.

Hayashi, K., *et al.* 1970. Ecological studies on Japanese encephalitis virus: results of investigations in the Nagasaki Area, Japan, in 1968 [in Japanese]. *Tropical Medicine* 11:212-220.

Hayes, R. O. 1962. Entomological aspects of the 1959 outbreak of eastern encephalitis in New Jersey. *American Journal of Tropical Medical Hygiene* 11:115-121.

Headlee, T. J. 1918. Migration as a factor in control. *Proceedings, New Jersey Mosquito Extermination Association* 5:104-112.

————. 1945. *The mosquitoes of New Jersey and their control*. Rutgers University Press, New Brunswick. 326 pp.

Hearle, E. 1926. The mosquitoes of the lower Frazer Valley, British Columbia and their control. *National Research Council Review, Ottawa*, No. 17. 94 pp.

————. 1932. Notes on the more important mosquitoes of western Canada. *Proceedings, New Jersey Mosquito Extermination Association* 19:7-14.

Hill, N. D. 1939. Biological and taxonomic observations on the mosquitoes of Kansas. *Transactions of the Kansas Academy of Science* 42:255-265.

Hodes, H. L. 1946. Experimental transmission of Japanese B encephalitis by mosquitoes and mosquito larvae. *Bulletin, Johns Hopkins Hospital* 79: 358-360.

Horsfall, W. R. 1937. Mosquitoes of southeastern Arkansas. *Journal of Economic Entomology* 30:743-748.

———. 1954. A migration of *Aedes vexans* Meigen. *Journal of Economic Entomology* 47:544.

———. 1956. Eggs of floodwater mosquitoes III (Diptera, Culicidae). Conditioning and hatching of *Aedes vexans. Annals of the Entomological Society of America* 49:66-71.

———. 1956a. A method of making a survey of floodwater mosquitoes. *Mosquito News* 16:66-71.

———. 1961. Traps for determining direction of flight of insects. *Mosquito News* 21:296-299.

———. 1962. Trap for separating collections of insects by interval. *Journal of Economic Entomology* 55:808-811.

———. 1963. Eggs of floodwater mosquitoes (Diptera: Culicidae) IX. Local distribution. *Annals of the Entomological Society of America* 56:426-441.

———. 1972. *Mosquitoes — their bionomics and relation to disease.* Hafner Publication Incorporated, New York. vii + 723 pp.

———, and G. B. Craig, Jr. 1956. Eggs of floodwater mosquitoes IV. Species of *Aedes* common in Illinois (Diptera: Culicidae). *Annals of the Entomological Society of America* 49:368-374.

Horsfall, W. R., L. M. Henderson, and P. T. M. Lum. 1957. Eggs of floodwater mosquitoes (Diptera: Culicidae) III. Effect of metabolites on latent embryos in uncapped eggs. *Proceedings of the Society for Experimental Biology and Medicine* 95:828-830.

Horsfall, W. R., P. T. M. Lum, and L. M. Henderson. 1958. Eggs of floodwater mosquitoes (Diptera: Culicidae) V. Effect of oxygen on hatching of intact eggs. *Annals of the Entomological Society of America* 51:209-213.

Horsfall, W. R., and M. L. Taylor. 1967. Temperature and age as factors in inducing insemination of mosquitoes (Diptera: Culicidae). *Annals of the Entomological Society of America* 60:118-120.

Horsfall, W. R., F. R. Voorhees, E. W. Cupp, and H. W. Fowler, Jr. 1969. Moment of hatching of *Aedes vexans* (Meigen): Feature photograph. *Annals of the Entomological Society of America* 62:253.

Horsfall, W. R., F. R. Voorhees, and E. W. Cupp. 1970. Eggs of floodwater mosquitoes. XIII. Chorionic sculpturing. *Annals of the Entomological Society of America* 63:1709-1716.

Hu, S. M. K. 1931. Studies on the host-parasite relationships of *Dirofilaria immitis* Leidy and its culicine intermediate hosts. *American Journal of Hygiene* 14:614-629.

Huffaker, C. B., and R. C. Bach. 1943. A study of the methods of sampling mosquito populations. *Journal of Economic Entomology* 36:561-569.

Hurlbut, H. S., and C. Nibley, Jr. 1964. Virus isolations from mosquitoes in Okinawa. *Journal of Medical Entomology* 1:78-83.

Iha, S. 1971. Feeding preference and seasonal distribution of mosquitoes in relation to the epidemiology of Japanese encephalitis in Okinawa Main Island [in Japanese]. *Tropical Medicine* 12:143-168.

Ikeshoji, T., and M. S. Mulla. 1970. Overcrowding factors of mosquito larvae 2. Growth-retarding and bacteriostatic effects of the overcrowding factors on mosquito larvae. *Journal of Economic Entomology* 63:1737-1743.

Irwin, W. H. 1941. A preliminary list of the Culicidae of Michigan. Part I. Culicinae (Diptera). *Entomology News* 52:101-105.

Jalil, M., and R. Mitchell. 1972. Parasitism of mosquitoes by water mites. *Journal of Medical Entomology* 9:305-311.

James, H. G. 1961. Some predators of *Aedes stimulans* (Walk.) and *Aedes trichurus* (Dyar) (Diptera: Culicidae) in woodland pools. *Canadian Journal of Zoology* 39:533-540.

Johnson, E. B. 1959. Distribution and relative abundance of mosquito species in Louisiana. *Technical Bulletin of the Louisiana Mosquito Control Association* No. 1. 18 pp.

Jolly, M., J. M. Klein, G. Audebaud, and J. J. Salaun. 1970. Une technique de précipitation en milieu gélifié pour l'identification des repas sanguins d'arthropodes. Application préliminaire aux moustiques de la région de Phnom-Penh (Cambodge). *Médecine tropicale* 30:236-247.

Jones, G., and J. R. Arnold. 1952. Report of the studies made at the College of the Pacific on embryological development of *Aedes nigromaculis* (Ludlow) and other floodwater mosquitoes. *Proceedings and Papers of the Annual Conference of the California Mosquito Control Association* 20:36-38.

Jones, J. C., and R. E. Wheeler. 1965. Studies on spermathecal filling in *Aedes aegypti* (Linnaeus). II. Experimental. *Biological Bulletin of the Marine Biological Laboratory*, Woods Hole, 129:532-545.

Joyce, C. R., and P. Y. Nakagawa. 1963. *Aedes vexans nocturnus* (Theobald) in Hawaii. *Proceedings, Hawaiian Entomological Society* 18:273-280.

Judd, W. W. 1954. Results of a survey of mosquitoes conducted at London, Ontario in 1952 with observations on the biology of the species collected. *Canadian Entomologist* 86:101-108.

Kardatzke, J. T., and K. K. Liem. 1972. Growth of *Aedes stimulans* and *A. vexans* (Diptera: Culicidae) in saline solutions. *Annals of the Entomological Society of America* 65:1425-1426.

Kato, M., and M. Toriumi. 1956. Emergence of larvae of *Aedes vexans* in the ground pool and the relations to the microorganism community [in Japanese with English summary]. *Ecological Review* 14:163-167.

Kennedy, J. S. 1939. The visual responses of flying mosquitoes. *Proceedings of the Zoological Society of London* (A) 109:221-242.

Khelevin, N. V. 1961. External factors influencing hatching of the larvae and number of generations of *Aedes vexans* Meig. (Diptera: Culicidae) [in Russian]. *Meditsinskaya parazitologiya i parazitarnÿe bolezni Moskva* 30:43-48.

King, W. V., G. H. Bradley, and T. E. McNeel. 1942. The mosquitoes of the southeastern states. *U.S. Department of Agriculture Miscellaneous Publication* No. 336. 96 pp.

King, W. V., G. H. Bradley, C. N. Smith, and W. C. McDuffie. 1960. A handbook of the mosquitoes of the southeastern United States. *Agriculture Handbook* No. 173, U.S. Department of Agriculture. 188 pp.

Klassen, W., and B. Hocking. 1964. The influence of a deep river valley system on the dispersal of *Aedes* mosquitoes. *Bulletin of Entomological Research* 55:289-304.

Kliewer, J. W. 1961. Weight and hatchability of *Aedes aegypti* eggs (Diptera: Culicidae). *Annals of the Entomological Society of America* 54:912-917.

Knab, F. 1907. Mosquitoes as flower visitors. *Journal of the New York Entomological Society* 15:215-219.

Knight, K. L. 1954. Mosquito light-trap collections at Yukon, Florida for the six-year period, 1948-1953. *Proceedings, New Jersey Mosquito Extermination Association* 41:251-257.

―――, and C. Henderson. 1967. Flight periodicity of *Aedes vexans* (Meigen) (Diptera: Culicidae). *Journal of the Georgia Entomological Society* 2:63-68.

Kurashige, Y. 1964. Ecological studies on mosquitoes. I. Ecology of mosquitoes in Tochigi Prefecture, Japan. *Bulletin of the Faculty of Liberal Arts, Utsunomiya University, Japan* 13:55-103.

Kuznetzov, V. G., and A. I. Mikheeva. 1970. Occurrence of the fungus *Coelomomyces* in larvae of *Aedes* from the Far East [in Russian]. *Parazitologiya* 4:392-393.

Laird, M. 1954. A mosquito survey in New Caledonia and the Belep Islands, with new locality records for two species of *Culex*. *Bulletin of Entomological Research* 45:285-293.

―――. 1956. Studies on mosquitoes and freshwater ecology in the South Pacific. *Bulletin of the Royal Society of New Zealand* #6. 213 pp.

Lake, R. W. 1967. Notes on the biology and distribution of some Delaware mosquitoes. *Mosquito News* 27:324-331.

Lea, A. O. 1970. Endocrinology of egg maturation in autogenous and anautogenous *Aedes taeniorhynchus*. *Journal of Insect Physiology* 16:1689-1696.

Lefkovitch, L. P., and R. A. Brust. 1968. Locating the eggs of *Aedes vexans* (Mg.) (Diptera: Culicidae). *Bulletin of Entomological Research* 58:119-122.

Love, G. J., R. B. Platt, and M. H. Goodwin, Jr. 1963. Observations on the spatial distribution of mosquitoes in southwestern Georgia. *Mosquito News* 23:13-22.

Love, G. J., and W. W. Smith. 1957. Preliminary observations on the relation of light-trap collections to mechanical sweep-net collections in sampling mosquito populations. *Mosquito News* 17:9-14.

MacCreary, D. 1941. Comparative density of mosquitoes at ground level and at an elevation of approximately one hundred feet. *Journal of Economic Entomology* 34:174-179.

―――, and L. A. Stearns. 1937. Mosquito migration across Delaware Bay. *Proceedings, New Jersey Mosquito Extermination Association* 24:188-197.

MacGregor, M. E. 1925. Mosquitoes under winter conditions. *Bulletin of Entomological Research* 15:357-358.

McDaniel, I. N., and W. R. Horsfall. 1957. Induced copulation of aedine mosquitoes. *Science* 125:745.

McGregor, T., and R. B. Eads. 1943. Mosquitoes of Texas. *Journal of Economic Entomology* 36:938-940.

McKiel, J. A., and A. S. West. 1961. Effects of repeated exposures of hypersensitive humans and laboratory rabbits to mosquito antigens. *Canadian Journal of Zoology* 39:597-603.

McLean, D. M. 1970. California encephalitis virus isolation from British Columbia mosquitoes. *Mosquito News* 30:144-145.

———, M. A. Crawford, S. R. Ladyman, R. R. Peers, and K. W. Purvin-Good. 1970. California encephalitis and Powassan virus activity in British Columbia, 1969. *American Journal of Epidemiology* 92:266-272.

McLintock, J. 1944. The mosquitoes of the greater Winnipeg area. *Canadian Entomologist* 76:89-104.

———, and A. N. Burton. 1967. Interepidemic hosts of western encephalitis virus in Saskatchewan. *Proceedings, New Jersey Mosquito Extermination Association* 54:97-104.

Mail, G. A. 1934. The mosquitoes of Montana. *Bulletin of the Montana Agricultural Experiment Station.* 288. 72 pp.

Malkova, D., V. Danielova, N. M. Kolman, J. Minar, and A. Smetna. 1965. Natural focus of Tahyna virus in South Moravia. *Journal of Hygiene, Epidemiology, Microbiology and Immunology* 9:434-440.

Marshall, J. F. 1938. *The British mosquitoes.* London: The British Museum. xi + 341 pp. + 20 pls.

Masters, C. O. 1962. A study of some biting habits of mosquitoes and their response to lights in a northern Ohio woods. *Mosquito News* 22:182-185.

Metcalf, R. L. 1945. The physiology of the salivary glands of *Anopheles quadrimaculatus*. *Journal of the National Malaria Society* 4:271-278.

Michener, C. D. 1945. Seasonal variations in certain species of mosquitoes (Diptera: Culicidae). *Journal of the New York Entomological Society* 53:293-300.

Miller, F. W. 1930. A progress report in an investigation of egg-laying habits of *Aedes sylvestris*. *Proceedings, New Jersey Mosquito Extermination Association* 17:105-111.

Minson, K. L. 1969. An *Aedes-vexans* gynandromorph. *Mosquito News* 29:135.

Mitchell, C. J., J. W. Kilpatrick, R. O. Hayes, and H. W. Curry. 1970. Effects of ultra-low volume applications of malathion in Hale County Texas II. Mosquito populations in treated and untreated areas. *Journal of Medical Entomology* 7:85-91.

Mitchell, E. G. 1907. *Mosquito life.* New York: G. P. Putnam Sons. xxii + 281 pp.

Möhrig, W. 1969. Die Culiciden Deutschlands. Untersuchungen zur Taxonomie, Biologie und Ökologie der einheimischen Stechmücken. *Parasitologische Schriftenriebe* No. 18. Gustav Fischer, Jena. 260 pp.

Morris, C. D., and G. R. De Foliart. 1971. Parous rates in Wisconsin mosquito populations. *Journal of Medical Entomology* 8:209-212.

Mukanov, S. M. 1970. Blood-sucking mosquitoes of the Udmurt ASSR [in Russian]. *Meditsinskaya parazitologiya i parazitarnye bolezni Moskva* 39:698-700.

Mulhern, T. D. 1936. A summary of mosquito-control accomplishments in 1935. *Proceedings, New Jersey Mosquito Extermination Association* 23: 14-51.

Myklebust, R. J. 1966. Distribution of mosquitoes and chaoborids in Washington State, by counties. *Mosquito News* 26:515-519.

Nayar, J. K., and D. M. Sauerman, Jr. 1968. Larval aggregation formation and population density interrelations in *Aedes taeniorhynchus*, their effects on pupal ecdysis and adult characteristics at emergence. *Entomologia Experimentalis et Applicata* 11:423-442.

————. 1970. A comparative study of grow and development in Florida mosquitoes. Part 1: Effects of environmental factors on ontogenetic timings, endogenous diurnal rhythm and synchrony of pupation and emergence. *Journal of Medical Entomology* 7:163-174.

Newton, W. L., W. H. Wright, and I. Pratt. 1945. Experiments to determine potential mosquito vectors of *Wuchereria bancrofti* in the continental United States. *American Journal of Tropical Medicine* 25:253-261.

Niebanck, L. 1957. Observations on mosquito species in Nassau County, Long Island, New York, based on larval and adult collections during the ten-year period, 1947-1956. *Mosquito News* 17:315-318.

Nielsen, E. T. 1958. The initial stage of migration in salt-marsh mosquitoes. *Bulletin of Entomological Research* 49:305-313 + 2 pl.

Nielsen, L. T., and D. M. Rees. 1961. An identification guide to the mosquitoes of Utah. *University of Utah Biology Series*, No. 12. v + 63 pp.

Novy, F. G., W. J. MacNeal, and H. N. Torrey. 1907. The trypanosomes of mosquitoes and other insects. *Journal of Infectious Diseases* 4:223-276 + 7 pls.

Olson, T. A., R. C. Kennedy, M. E. Reuger, R. D. Price, and L. L. Schlottman. 1961. Evaluation of activity of viral encephalitides in Minnesota through measurement of pigeon antibody response. *American Journal of Tropical Medical Hygiene* 10:266-270.

Owen, W. B. 1937. The mosquitoes of Minnesota, with special reference to their biologies. *Technical Bulletin of the Minnesota Agricultural Experiment Station* 126. 75 pp.

————, and R. W. Gerhardt. 1957. The mosquitoes of Wyoming. *University of Wyoming Publications* 21:71-141.

Paulovova, J. 1967. Notes on the biometrics of imagines of *Aedes vexans* Meig. from environs of Hodonin during 1966. *Biologia, Bratislava* 22:525-528.

Peterson, A. G., and W. W. Smith. 1945. Occurrence and distribution of mosquitoes in Mississippi. *Journal of Economic Entomology* 38:378-383.

Philip, C. B. 1943. Flowers as a suggested source of mosquitoes during encephalitis studies, and incidental mosquito records in the Dakotas in 1941. *Journal of Parasitology* 29:328-329.

Porter, C. H., and W. L. Gojmerac. 1969. Field observations with abate and bromophos: their effect on mosquitoes and aquatic arthropods in a Wisconsin park. *Mosquito News* 29:617-620.

Post, R. L., and J. A. Munro. 1949. Mosquitoes of North Dakota. *North Dakota Bimonthly Bulletin* 11:173-183.

Price, R. D. 1960. Identification of first-instar aedine mosquito larvae of Minnesota (Diptera: Culicidae). *Canadian Entomologist* 92:544-560.

———. 1963. Frequency of occurrence of spring *Aedes* (Diptera: Culicidae) in selected habitats in northern Minnesota. *Mosquito News* 23:324-329.

———. 1964. Mosquito abundance within a Minnesota metropolitan area, 1956-1962. *Canadian Entomologist* 96:1034-1036.

Provost, M. W. 1960. The dispersal of *Aedes taeniorhynchus* III. Study methods for migratory exodus. *Mosquito News* 20:148-161.

Pucat, A. 1965. List of mosquito records from Alberta. *Mosquito News* 25:300-302.

Quinby, G. E., R. E. Serfling, and J. K. Neel. 1944. Distribution and prevalence of the mosquitoes of Kentucky. *Journal of Economic Entomology* 37:547-550.

Quraishi, M. S., R. A. Brust, and L. P. Lefkovitch. 1966. Uptake, transfer and loss of p^{32} during metamorphosis, mating and oviposition in *Aedes vexans*. *Journal of Economic Entomology* 59:1331-1333.

Rees, D. M. 1943. The mosquitoes of Utah. *Bulletin, University of Utah Biology Series*, No. 33. 99 pp.

———, and F. C. Harmston. 1948. Mosquito records from Wyoming and Yellowstone National Park. (Diptera: Culicidae). *Pan-Pacific Entomologist* 24:181-188.

Reeves, W. C., and W. M. Hammon. 1944. Feeding habits of the proven and possible vectors of western equine and St. Louis encephalitis in the Yakima Valley, Washington. *American Journal of Tropical Medicine* 24:131-134.

Reeves, W. C., and A. Rudnick. 1951. A survey of the mosquitoes of Guam in two periods in 1948 and 1949 and its epidemiological implications. *American Journal of Tropical Medicine* 31:633-658.

Reisen, W. K., J. P. Burns, and R. G. Basio. 1972. A mosquito survey of Guam, Marianas Islands, with notes on the vector-borne disease potential. *Journal of Medical Entomology* 9:305-311.

Richards, A. G. 1938. Mosquitoes and mosquito control on Long Island, New York, with particular reference to the salt-marsh problem. *Bulletin of the New York State Museum* 316:85-180.

Richards, C. S., L. T. Nielsen, and D. M. Rees. 1956. Mosquito records from the Great Basin and the drainage of the Lower Colorado River. *Mosquito News* 16:10-17.

Roberts, R. H., W. A. Pond, H. F. McCrory, J. W. Scales, and J. C. Collins. 1969. Culicidae and Tabanidae as potential vectors of anaplasmosis in Mississippi. *Annals of the Entomological Society of America* 62:863-868.

Rösch — . 1933. Die *Aedes vexans* — Plage im Mainzer Becken und die Rheingelstande. *Zeitschrift für Gesundheitstechnik und Städtehygiene* 25:629-636.

Ross, H. H. 1947. The mosquitoes of Illinois (Diptera, Culicidae). *Bulletin of the Illinois State Laboratory of Natural History*, No. 24. 96 pp.

———, and W. R. Horsfall. 1965. A synopsis of mosquitoes of Illinois (Diptera, Culicidae). *Illinois Natural History Survey Biological Notes*, No. 52. 50 pp.

Rowe, J. A. 1942. Mosquito light-trap catches from ten Iowa cities, 1940. *Iowa State College Journal of Science* 16:487-518.

Rozeboom, L. E. 1942. The mosquitoes of Oklahoma. *Technical Bulletin of the Oklahoma Agricultural Experiment Station,* No. 16. 56 pp.

Rudolfs, W., and J. B. Lackey. 1929. Effect of food upon phototropism of mosquito larvae. *American Journal of Hygiene* 10:245-252.

Sasa, M. 1964. Observation on the bionomics of mosquitoes in the Amami and Tokyo areas by indoor human-bait collections with special reference to their nocturnal biting rhythm. *Japanese Journal of Experimental Medicine* 34:89-107.

Scanlon, J. E., and S. Essah. 1965. Distribution in altitude of mosquitoes in northern Thailand. *Mosquito News* 25:137-144.

Scherer, W. F., M. Funkenbusch, F. L. Buescher, and T. Izumi. 1962. Sagiyama virus, a new Group-A arthropod-borne virus from Japan. I. Isolation, immunologic classification, and ecologic observations. *American Journal of Tropical Medical Hygiene* 11:255-268.

Schlaifer, A., and D. E. Harding. 1946. The mosquitoes of Tennessee. *Journal of the Tennessee Academy of Science* 21:241-256.

Seaman, E. A. 1945. Ecological observations and recent records on mosquitoes of San Diego and Imperial counties, California. *Mosquito News* 5:89-95.

Shemanchuk, J. A. 1969. Epidemiology of western encephalitis in Alberta: response of natural populations of mosquitoes to avian host. *Journal of Medical Entomology* 6:269-275.

Shichijo, A., K. Mifune, K. Hayashi, Y. Wada, S. Ito, S. Kawai, I. Miyagi, and T. Oda. 1968. Ecological studies on Japanese encephalitis virus. Survey of virus dissemination in Nagasaki area, Japan, in 1966 and 1967. *Tropical Medicine* 10:168-180.

Shichijo, A., *et al.* 1970. Isolation of Japanese encephalitis virus and group A arboviruses from *Aedes vexans nipponii* caught in the Nagasaki area, Japan [in Japanese]. *Tropical Medicine* 12:91-97.

Simkova, A., V. Danielova, and V. Bardos. 1960. Experimental transmission of the Tahyna virus by *Aedes vexans* mosquitoes. *Acta Virologica,* Prague, 4:341-347.

Siverly, R. E. 1966. Mosquitoes of Delaware County, Indiana. *Mosquito News* 26:221-229.

Smith, J. B. 1903. Concerning mosquito migrations. *Science* 18:761-764.

———. 1904. Report on the mosquito investigation. *Report of the Entomology Department, New Jersey Agricultural Experiment Station,* 1903, 643-659.

———. 1904a. Report upon the mosquitoes occurring within the state, their habits, life history, etc. *Report of the Entomology Department, New Jersey Agricultural Experiment Station,* 1904. 482 pp.

———. 1904b. Notes on some mosquito larvae found in New Jersey. *Entomological News* 15:145-154 + 4 pls.

Smith, W. W., and G. J. Love. 1956. Effects of drought on the composition of rural mosquito populations as reflected by light-trap catches. *Mosquito News* 16:279-281.

Sommerman, K. M. 1968. Notes on Alaskan mosquito records. *Mosquito News* 28:233-234.

Spielman, A. 1964. The mechanics of copulation in *Aedes aegypti*. *Biological Bulletin of the Marine Biological Laboratory* 127:324-344.

Stabler, R. M. 1945. New Jersey light-trap versus human bait as a mosquito sampler. *Entomological News* 56:93-99.

———. 1952. Parasitism of mosquito larvae by mermithids (Nematoda). *Journal of Parasitology* 38:130-132.

Stage, H. H., C. M. Gjullin, and W. W. Yates. 1937. Flight range and longevity of floodwater mosquitoes in the lower Columbia River Valley. *Journal of Economic Entomology* 30:940-945.

———. 1952. Mosquitoes of the northwestern states. *U.S. Department of Agriculture Handbook* No. 46. 95 pp. + 1 pl.

Stage, H. H., and W. W. Yates. 1936. Some observations on the amount of blood engorged by mosquitoes. *Journal of Parasitology* 22:298-300.

Stamm, D. D. 1968. Arbovirus studies in birds in south Alabama, 1959-1960. *American Journal of Epidemiology* 87:127-137.

Steiner, G. 1924. Remarks on a mermithid found parasitic in the adult mosquito (*Aedes vexans* Meigen) in B.C. *Canadian Entomologist* 56:161-164.

Steward, C. C., and J. W. McWade. 1961. The mosquitoes of Ontario (Diptera: Culicidae) with keys to the species and notes on distribution. *Proceedings of the Entomological Society of Ontario* 91:121-188.

Stone, A., K. L. Knight, and H. Starcke. 1959. A synoptic catalog of the mosquitoes of the world (Diptera: Culicidae). *Entomological Society of America*. 358 pp.

Strong, L. A. 1904. *Report of the Bureau of Entomology and Plant Quarantine, U.S. Dept. of Agriculture*. 1939. 40:94.

Sudia, W. D., R. W. Chamberlain, and P. H. Coleman. 1968. Arbovirus isolations from mosquitoes collected in south Alabama, 1959-1963, and serologic evidence of human infection. *American Journal of Epidemiology* 87:112-126.

Sudia, W. D., V. F. Newhouse, C. H. Calisher, and R. W. Chamberlain. 1971. California-group arboviruses: isolations from mosquitoes in North America. *Mosquito News* 31:576-600.

Summers, W. A. 1943. Experimental studies on the larval development of *Dirofilaria immitis* in certain insects. *American Journal of Hygiene* 37:173-178.

Sutherland, D. J., and L. E. Hagmann. 1963. Larval resistance studies in New Jersey 1962. *Proceedings, New Jersey Mosquito Extermination Association* 50:297-302.

Sweet, B. H., and J. S. McHale. 1970. Characterization of cell lines from *Culiseta inornata* and *Aedes vexans* mosquitoes. *Experimental Cell Research* 61:51-63.

Symons, T. B., T. H. Coffin, and A. B. Gahan. 1906. The mosquito. *Bulletin of the Maryland Agricultural Experiment Station* 109:73-124.

Tate, H. D., and D. B. Gates. 1944. The mosquitoes of Nebraska. *Research Bulletin, Nebraska Agricultural Experiment Station* 133:1-27.

Tempelis, C. H., R. O. Hayes, A. D. Hess, and W. C. Reeves. 1970. Blood-feeding habits of four species of mosquito found in Hawaii. *American Journal of Tropical Medical Hygiene* 19:335-341.

Ten Broeck, C., and M. H. Merrill. 1935. Transmission of equine encephalomyelitis by mosquitoes. *American Journal of Pathology* 11:847.

Thompson, P. H. 1964. The biology of *Aedes vexans* in Wisconsin (Culicidae, Diptera). Ph.D. thesis. Madison: University of Wisconsin.

———, and R. J. Dicke. 1965. Sampling studies with *Aedes vexans* and some other Wisconsin *Aedes* (Diptera: Culicidae). *Annals of the Entomological Society of America* 58:927-930.

Trpis, M. 1962. Ökologische Analyse der Stechmückenpopulationen in der Donautiefebene in der Tschechoslowakei. *Biologia, Bratislava* 8(3):132 pp.

———, W. O. Haufe, and J. A. Shemanchuk. 1968. Mermithid parasites of the mosquito *Aedes vexans* Meigen in British Columbia. *Canadian Journal of Zoology* 46:1077-1079.

Trpis, M., and J. A. Shemanchuk. 1970. Effect of constant temperature on the larval development of *Aedes vexans* (Diptera: Culicidae). *Canadian Entomologist* 102:1048-1051.

Tshinaev, P. P. 1945. Flying activity and attack on man of various species of Anopheles and culicines under natural conditions in Uzbekistan [in Russian]. *Meditsinskaya parazitologiya i parazitarnÿe bolezni Moskva* 14:15-35.

Twinn, C. R. 1931. Notes on the biology of mosquitoes of eastern Canada. *Proceedings, New Jersey Mosquito Extermination Association* 18:10-22.

———. 1931a. Observations on some animal and plant enemies of mosquitoes. *Canadian Entomologist* 63:51-61.

———. 1949. Mosquitoes and mosquito control in Canada. *Mosquito News* 9:35-41.

———. 1953. Mosquitoes and their control in Prince Edward Island. *Mosquito News* 13:185-190.

Venard, C. E., and F. W. Mead. 1953. An annotated list of Ohio mosquitoes. *Ohio Journal of Science* 53:327-333.

Wallis, R. C., R. M. Taylor, and J. R. Henderson. 1960. Isolation of eastern equine encephalitis virus from *Aedes vexans* in Connecticut. *Proceedings of the Society of Experimental Biology* 103:442-444.

Wesenberg-Lund, C. 1921. Contributions to the biology of the Danish Culicidae. *Det Kongelige Danske Videnskabernes Selskabes Skrifter* (Ser. 8), 7:210 pp.

Wilson, C. A., R. C. Barnes, and H. L. Fellton. 1946. A list of mosquitoes of Pennsylvania with notes on their distribution and abundance. *Mosquito News* 6:78-84.

Wilson, G. R., and W. R. Horsfall. 1970. Eggs of floodwater mosquitoes. XII. Installment hatching of *Aedes vexans* (Diptera: Culicidae). *Annals of the Entomological Society of America* 63:1644-1647.

Wong, Y. W., J. A. Rowe, M. C. Dorsey, M. J. Humphreys, and W. J. Hausler, Jr. 1971. Arboviruses isolated from mosquitoes collected in southeastern Iowa in 1966. *American Journal of Tropical Medical Hygiene* 20:726-729.

Woodard, D. B., and H. C. Chapman. 1965. Blood volumes ingested by various pest mosquitoes. *Mosquito News* 25:490-491.

Wright, R. E., and K. L. Knight. 1966. Effect of environmental factors on biting activity of *Aedes vexans* (Meigen) and *Aedes trivittatus* (Coquillett). *Mosquito News* 26:565-578.

Yen, C. H. 1938. Studies on *Dirofilaria immitis* Leidy, with special reference to the susceptibility of some Minnesota species of mosquitoes to infection. *Journal of Parasitology* 24:189-205.

EMBRYOLOGY

Louis J. Moretti and Joseph R. Larsen

INTRODUCTION

Work in insect embryology began in the early nineteenth century; a large body of contributions came from German investigators. All of these earlier investigations have been compiled, summarized, and compared by Korschelt and Heider (1899) and may be found in volume III of their four-volume series on invertebrate embryology. Up until this study descriptive embryological works dealt primarily with single organisms and were not comparative in nature. A more recent summarization of the literature on insect development is a textbook by Johannsen and Butt (1941). The most recent contribution is a text edited by Kumé and Katsuma (1968) on invertebrate embryology. This work devotes a section to insect embryology and includes the major contributions of Japanese investigators which heretofore have not been translated.

Of any single dipterous insect, the most detailed body of embryological information exists for *Drosophila melanogaster*. The early stages of development of this organism are described by Sonnenblick (1950) and the later stages by Poulson (1950), based on an earlier work by the same author (Poulson, 1937). Fish (1947a, 1947b, 1949, 1952) has studied the early cleavage, blastoderm, inner layer, and gut formation of *Phaenicia sericata*.

More recently Anderson (1961, 1962a, 1962b, 1963, 1964) has published a series of papers on the embryonic development of *Dacus tryoni*, the Queensland fruit fly. These works are detailed and informative and, unlike most other studies, give considerable attention to the location and fates of the imaginal primordia.

Embryological studies on the Culicidae have only recently appeared in the literature. Ivanova-Kazas (1947) published a paper on the embryonic development of *Anopheles maculipennis*. This work was accomplished primarily through the study of serially sectioned material. The author provides a detailed account of the embryogeny and gives particular emphasis to the process of segmentation. Organogenesis, however, is treated briefly. Nonetheless, this probably remains as one of the two most complete works on mosquito embryology.

In an unpublished thesis Larsen (1952) presents a histological study of the embryology of *Culiseta inornata*. This work provides information on the early stages of development and traces the origins of the major organ systems. Adequate descriptions of pre-cleavage phenomena and gonadal development are lacking; also, the descriptions of organogenesis are somewhat superficial. An additional study on the early embryology of *Culiseta inornata* has recently been provided by Harber (1969) and Harber and Mutchmor (1970). Standard histological as well as histochemical techniques are employed in this investigation.

The gross embryology of three species of aedine mosquitoes has been described by Telford (1957). This author devotes particular attention to mouthpart development and segmentation. A more superficial study on the gross external morphology of *Culex tarsalis* embryos is provided by Rosay (1959). In his book on the life history of *Aedes aegypti*, Christophers (1960) describes the external changes taking place in the development of *Culex molestus*.

Unquestionably the most outstanding work on mosquito embryology to date is that of Idris (1960) on *Culex pipiens*. This investigation provides an extremely detailed and well illustrated analysis of all events leading to organogenesis. It is based on a study of serial sections and whole mount preparations.

Davis (1967) studied the embryonic development of *Culex fatigans* and compared it to that of *Lucilia sericata*. This histological study presents a concise description of preblastoderm development, blastula formation, germ band formation, and organogenesis. Also of interest is the construction of a fate map showing the potentialities of the various regions of the blastoderm.

The most recent contribution to the field of mosquito embryology is a gross and histological study by Guichard (1971) on *Culex pipiens* and *Calliphora erythrocephala*. The primary emphasis of this work is given to segmentation and development of the mouthparts.

It should be noted that the more detailed embryological studies are limited to the genuses *Culex* and *Anopheles*. Virtually nothing is known about development in the genus *Aedes*. This stems from the fact that these are floodwater mosquitoes and oviposit their eggs in areas that are dry during part of the egg stage. As one of the measures to prevent dessication, these eggs form a hard, brittle endochorion. This structure, in addition to the underlying impermeable vitelline membrane, creates a problem for the penetration of fixing and embedding compounds and further for the ability to obtain satisfactory serial sections.

Members of this genus such as *Aedes aegypti* have proved to be of great importance as vectors of disease; knowledge of their embryonic stage could have control applications. Information on development in this group would be of value to the comparative embryologist and phylogeneticist, as well as filling a gap in our present knowledge of culicid embryology. With these reasons in mind the present study was undertaken on *Aedes vexans*. The intent is to provide a concise description of preblastoderm, blastoderm, and germ band formation followed by organogenesis.

MATERIALS AND METHODS

Adult female *Aedes vexans* were collected in Champaign County during the summer of 1969. The mosquitoes were allowed to alight on the collector and were then captured in a plastic collecting tube and transferred to a rectangular oviposition cage (Abou-Aly, 1968). When returned to the laboratory, the mosquitoes were permitted to take a blood meal from the collector. The cages were then placed over a layer of cellucotton and muslin which was kept saturated with deionized water. The adults were stored in this manner for four days at room temperature under constant light. At the end of the fourth day oviposition was induced by decapitation (DeCoursey and Webster, 1952). Oviposition was allowed to proceed on moist filter paper for ten minutes, and at the end of this time the female was removed. The filter paper containing the eggs was then transferred to a small disposable petri dish lined with wet cellucotton. The petri dish was then covered, labeled, and placed in a water bath incubator maintaining a temperature of 25°C with a fluctuation of less than 0.10°. This procedure was repeated on three separate females for each stage studied.

Eggs were fixed at one-hour intervals from 0 to 16 hours, 18½ hours, and five-hour intervals between 21 and 96 hours. In this text the number of hours of incubation designates a developmental stage. After the specified incubation time the petri dishes were removed and held at 4°. This served to stop embryonic development so that these eggs could be worked on for several hours without disturbing the level of development within a particular stage. Examination of sectioned material showed this method to be highly effective. Small groups of eggs were removed from each petri dish and placed in a depression slide containing a bleaching compound composed of 2.50 gms of sodium chlorite, 25 ml of acetic acid, and 300 ml of distilled water (modified from Trpis, 1970). Exposure to this solution for two minutes was sufficient to remove the exochorion and soften and bleach the endochorion. This step was necessary to give the egg the proper consistency for the subsequent histological procedures. The eggs were then pipetted into a small, wax-bottomed dish filled with Huettner's modification of Kahle's fixative (Demerec, 1950).

At incubation time less than 18½ hours, either the anterior or posterior tenth of the egg was removed with a tungsten needle under a dissecting microscope. The eggs were then transferred to a vial of fresh fixative, where they remained for 24 hours, followed by storage in 70 percent ethanol. When incubation times were 18½ hours or older, the anterior seventh of the endochorion was carefully peeled away with a tungsten needle. The underlying vitelline membrane was also removed; its removal is an essential step for satisfactory penetration of fixative and embedding medium.

The eggs were dehydrated in a graded series of alcohols and embedded in a mixture of methyl- and butylmethacrylate according to the following method.

1. Fix for 30 minutes to two hours in 3-5 ml of fixative in the cold (ice bath or cold room). Shake bottles gently two or three times during the period of fixation.

2. Rinse in half-strength buffer for five minutes. Cold.

3. Fifteen minutes in 25 percent ethyl alcohol. Cold.

4. Fifteen minutes in 50 percent ethyl alcohol. Cold.

5. Fifteen minutes in 75 percent ethyl alcohol. Cold.

6. Thirty minutes in 95 percent ethyl alcohol. Allow the specimens to come to room temperature in this solution. All following steps are done at room temperature unless otherwise specified.

7. Thirty minutes in absolute alcohol.

8. Thirty minutes in a fresh change of absolute ethyl alcohol.

9. Thirty minutes in a fresh change of absolute ethyl alcohol. Procedure for embedding in the mixture of butyl- and methylmethacrylate.

10. Thirty minutes in 50-50 absolute alcohol and a mixture of butyl-

and methylmethacrylate using one part butyl to three parts methyl-methacrylate.

11. Thirty minutes in 1-3 mixture of butyl- and methylmethacrylate.

12. Thirty minutes in 1-3 mixture of butyl- and methylmethacrylate plus catalyst.

13. Embed in fresh 1-3 mixture butyl- and methylmethacrylate which has been pre-polymerized with a catalyst. The catalyst used for poly-merization was a 2,4-dichlorobenzoil peroxide in dibutyl phthalate, which is sold under the trade name "luberco CDB." The catalyst was added to the mixture 1-3 of butyl and methylmethacrylate at room temperature in the amount of 2 gms per 100 cc.

Serial sections were cut at a thickness of 0.80 to 1.20 μ on a Porter Blum MT-2 ultramicrotome fitted with a Dupont diamond knife. The sections were generally stained with Delafield's hematoxylin, although with some older stages Heidenhain's iron hematoxylin was used. Photomicrographs were taken with a Leitz Ortholux microscope equipped with an Orthomat automatic camera.

A series of embryos preserved at stages 21 through 96 was pre-pared as whole mounts by entirely removing the shell, staining briefly in aceto-orcein, and mounting in diaphane.

RESULTS

Early Development

EGG STRUCTURE

At the time of oviposition the eggs are pearly white. By the end of the first hour they have taken on a purplish-gray color, which over the next two hours changes to a dark gray. By the end of the fourth or fifth hour the shell has blackened and the contents are completely obscured from view. The egg measures approximately 0.50 mm x 0.13 mm; it is decidedly blunt anteriorly and tapers toward the posterior pole. Its dorsal surface is usually flat to slightly concave, whereas the ventral surface is convex (Fig. 1).

The Law of Hallez (Hallez, 1886) states that at oviposition the three axes of the egg are oriented in the same manner as the axes of the maternal organism; further, the different surfaces of the egg coincide with the corresponding surfaces of the embryo. This law does not apply rigidly to the eggs of *Aedes vexans*, since as the egg passes out of the ovipositor only its anterior and posterior orientation consistently correspond to those of the mother; in addition, before blastokinesis occurs, the venter of the embryo develops on the dorsum of the egg.

Internally at stage 0 the egg is composed of an extensive yolk matrix or deutoplasm which in many areas lies in direct contact with the endochorion. Yolk spheres can be characterized as large or small. The large spheres have an average diameter of 9.6 μ; the small, 2.4 μ. The small spheres are located primarily at the periphery of the egg, whereas the larger ones are found in the interior (Fig. 2).

An interconnecting cytoplasmic network is found between the yolk spheres and extends to the egg surface (Fig. 3). A distinct periplasm was not observed, although noticeable concentrations of cytoplasm were present at both poles. A vitelline membrane is not present at this stage.

At the anterior pole a micropylar area is readily distinguishable. It consists of a single spherical opening in the egg shell which at this time is sealed by a dense plug composed primarily of nuclei (Fig. 4).

During its passage through the common oviduct usually one or two sperm enter the egg and at oviposition are found lying in a cytoplasmic island within the deutoplasm at the equatorial region. The sperm heads appear as slightly elongate dense masses of chromatin (Fig. 5). Apparently the tails are lost shortly after entry into the egg.

A dense mass of lightly staining granules is found in the posterior pole plasm; these polar granules measure about 0.90 μ in diameter and may be either arranged as a convex disc or scattered (Figs. 2, 3).

EARLY CLEAVAGE AND BLASTODERM FORMATION

The egg nucleus is not present initially but condenses during the first half hour of development. It undergoes meiosis I (Fig. 6) and produces the first polar body at the equatorial surface. The female pronucleus then returns to the interior of the deutoplasm, while the polar body undergoes meiosis II (Fig. 7); the two polar bodies remain at the surface, where they degenerate within a few hours. By the end of stage 2 the sperm head has transformed itself into the male pronucleus and has traveled anteriorly with the female pronucleus, where they fuse (Fig. 8). These pronuclei possess a clear nucleoplasm marbled with strands of strongly staining chromatin (Fig. 8).

Between stages 2 and 3 two successive mitotic divisions have occurred, producing four nuclei. However, supernumerary sperm may have also undergone the transformation into male pronuclei, since it was not uncommon to find eggs containing five to ten nuclei. By stage 4 up to 95 nuclei are present; based on these observations, the average mitotic cycle lasts about twenty minutes. In all ensuing stages its duration increases. This nuclear count also demonstrates that between stages 3 and 4 mitotic asynchrony has taken place. This asynchrony was also detectable cytologically by the presence of cleavage nuclei in different stages of the mitotic cycle. The cleavage nuclei are scattered throughout the deutoplasm, each within a cytoplasmic island, but do not appear closer than 14 μ from the egg surface. They are usually spherical and have a diameter of approximately 9 μ (Fig. 9). The chromatin is concentrated around the nuclear periphery, but some is found scattered throughout the center. The nucleoplasm has a dense, grayish appearance. By stage 6 many of the cleavage nuclei, with their surrounding cytoplasm, have migrated to the egg periphery.

Stages 7, 8, 9, and 10 are similar in that all of the cleavage nuclei have reached the egg surface to form the blastoderm (Fig. 10). The blastodermal nuclei are spherical to oval, anucleolate, and without cell boundaries. These stages differ in the number of nuclei which are present. The nuclei are embedded in a matrix of cytoplasm which often appears thicker on the side adjacent to the eggshell. This cytoplasm takes its origin from the cytoplasm which surrounded the cleavage nuclei before peripheral migration.

Between stages 10 and 11 the first observable difference between the cells that form the germ band/amnion and those that will give rise to the serosa were noted. In two specimens all the nuclei of the dorsal and lateral blastoderm were undergoing mitosis, while the nuclei of the ventral blastoderm were in the interphase stage (Fig. 11). Other noteworthy changes at this time are the appearance of a nucleolus in some of the blastodermal nuclei and of slight peripheral cytoplasmic indentations between adjacent nuclei. These indentations are often obscured by the closely adherent endochorion. The inner cytoplasmic surface is uneven, probably as a result of mechanical pressure by the yolk spheres.

At stage 13 the blastoderm is composed of a cuboidal epithelium with an uneven surface facing the yolk matrix. The nuclei of the dorsal blastoderm are centrally located in the cuboid mass of cytoplasm, whereas those of the ventral blastoderm are closer to the periphery. Lateral and peripheral cell boundaries are present, but they are absent centrally. Each nucleus contains a single nucleolus. At fourteen hours the cell boundaries are complete on all sides.

POLE CELLS

The pole or primordial germ cells can be traced easily in *Aedes vexans* because of their distinct cytological appearance. They first appear at the beginning of the fifth hour and are present at the posterior pole. They are approximately 12.7 μ in diameter with noticeably larger nuclei (Fig. 12). Their nucleoplasm is lighter and more vesicular than that found in cleavage nuclei; the chromatin occurs primarily in the central portion of the nucleus in the form of long, scattered rods. Some of these nuclei possess a cytoplasmic membrane and are true pole cells. These cells appear fragile and plastic, with their shape conforming to the adjacent area. Many of the polar granules at this stage are found in the cytoplasm of these cells.

By stage 6 ten pole cells are usually present and have taken a position at the posterior pole lying outside, but in contact with, the blastoderm (Fig. 13). They are irregular in shape, with a distinct nuclear membrane. The chromatin is usually condensed into a peripheral ring which is surrounded by a clear layer of nucleoplasm; a nucleolus is present. The cytoplasm has a somewhat reticular appearance and contains large numbers of polar granules, and not infrequently a few small yolk spheres. The definitive number of pole cells is about eleven. These cells are derived only from cleavage nuclei entering the polar region and never as a mitotic product of another pole cell. From stage 7 to stage 15 these cells remain in the same position just outside of the blastoderm. They gradually lose their fragile, irregular appearance and take on a more definite spherical form. At stage 11 the polar granules in many specimens were beginning to coalesce so that several large granules were present rather than many small ones (Fig. 14).

YOLK CELLS

The yolk cells or vitellophages function in the breakdown of yolk spheres and facilitate the availability of nutrients to the developing embryo. In conformity with most of the previous literature, vitellophages will be referred to as true cells even though they never possess a membrane surrounding their ameboid-like cytoplasm.

Yolk cells first appear at stage 5 simultaneously with the pole cells. Their nuclei are much smaller than the cleavage nuclei and usually appear elongate, but they may take on a variety of shapes as a result of pressure exerted by surrounding yolk spheres. The chromatin appears either as a highly compacted ring in the center of the nucleus or as a central granular mass (Fig. 15). At stage 6 some yolk cells appear to have migrated into the blastoderm and can easily be identified by their distinct cytological characteristics as described above (Fig. 16). Soon their appearance changes; they become indistinguishable from other blastodermal nuclei.

From stage 7 to stage 13 the yolk cells remain essentially unchanged except that the chromatin is more commonly in the form of a small,

dense mass surrounded by a thin, clear nucleoplasm. There appeared
to be an increased number of yolk cells in the deutoplasm at stage 13.
The nuclei of many of these cells appeared ameboid, with scattered
granules of chromatin (Fig. 17). It is not uncommon to find two nuclei
within the same unit of cytoplasm forming a syncytium. Yolk cells
found in contact with the blastoderm were interpreted as being secondary vitellophages which were migrating from the blastoderm into
the deutoplasm.

During the remainder of embryogeny the vitellophages consist of
nuclei enclosed by very little cytoplasm. They are usually scattered
throughout the yolk matrix, although in a few specimens they appeared
to be more highly concentrated at the yolk periphery. However, these
exceptional cases seemed insufficient to suggest the existence of a yolk
sac. At stages 91 to 96 some yolk cell nuclei appeared pycnotic, though
at the time of hatching many appeared to be normal. The yolk matrix
also undergoes certain changes during development. Beginning at
stage 46 and continuing until hatching, the yolk spheres in many
areas coalesce to form large homogeneous masses. Also, the yolk material found in contact with embryonic tissue is usually highly
vacuolated.

GERM BAND FORMATION AND GASTRULATION

Between stages 14 and 15 the cells of the dorsal and ventral blastoderm range in form from columnar to cuboidal. By stage 15 the formation of the germ band has become evident dorsally. The dorsal
blastoderm is composed of columnar cells with elongate nuclei (Fig.
18). The cells of the dorso-lateral blastoderm are also columnar, but
they decrease in height ventrad until the cells are distinctly flattened
within the ventral blastoderm with their oval to elongate nuclei lying
parallel to the surface of the egg (Fig. 18). It is this dorsal columnar
germ band that will give rise to the embryo proper, while its lateral
edges will form the amnion. The remaining cells of the lateral and
ventral blastoderm will give rise to the serosa.

At stage 16 the inner cytoplasmic surface of the cells has become
highly vacuolated, a condition which is probably indicative of an increased metabolic activity of these cells.

At this stage of development the pole cells migrate anteriorly through
the blastoderm (Figs. 14 and 19). They are followed in their migration
by blastodermal cells, which soon form a tongue-like structure extending anteriorly; this tongue of cells is the ventral extension of the germ
band. The pole cells are positioned in a clump at the dorsal tip of this
extension.

At stage 18½ the germ band begins at the anterior pole, extends dorsally along the full length of the egg, bends ventrally at the posterior pole, and proceeds along the mid-ventral margin of the egg to a point halfway between the anterior and posterior pole. The pole cells still remain at the dorsally extending tip of the germ band (Fig. 20).

A finger-like outpouching of cells occurs at each lateral margin of the germ band and grows dorsally to meet at the midline. They partially fuse along the anterior two-thirds of the dorsal and the anterior one-half of the ventral germ band, producing two distinct cellular layers (Figs. 21, 22). The layer of cells immediately covering the germ band is the amnion; the space between the germ band and this membrane is the amniotic cavity. The outer cell layer is the serosa, which envelops the entire egg contents. It is derived from the lateral and ventral blastoderm, and during the process of overgrowth its cells become highly attenuated. The cells of the amnion are still quite large.

By stage 18½ the prospective mesodermal cells which lie along the mid-line region of the germ band have begun to migrate ventrally; their migration is more advanced anteriorly. Once below the ectoderm the mesodermal cells arrange themselves into a solid cord, which shortly thereafter flattens to form a two-celled layer (Fig. 23). The mesodermal layer is usually completed by stage 31. It lies directly beneath and is adherent to the ectoderm. There is no gastrular furrow through which the mesodermal cells move; rather, the prospective mesodermal cells of the germ band simply elongate and migrate ventrally. Toward the end of gastrulation a mid-dorsal groove does appear, but this is associated with the development of the ventral nerve cord.

EMBRYONIC BODY FORM AND SEGMENTATION

At 21 hours of development the ventral extension of the germ band has reached a point three-fourths of the way to the anterior pole (Fig. 24A). Although *in toto* preparations reveal the presence of only a few indistinct segments, longitudinal sections demonstrate seventeen well-defined segments. Each segment is distinguishable as a slight indentation of the surface of the germ band and a correspondingly pronounced constriction of the mesodermal layer (Figs. 24B, 25). The most anterior or pregnathal segment will eventually develop into the labrum, antennae, and cephalon; this brings the total number of segments to nineteen. This is followed by the three gnathal (mandibular, maxillary, and labial), three thoracic, and ten abdominal segments (Fig. 24C). Segments 7-19 usually measure 20 μ in length each.

Up to stage 36 the major portion of the germ band has been developing against the dorsal surface of the egg; however, between stages

36 and 46 blastokinesis takes place. This process involves a 180° rotation of the embryo about its longitudinal axis so that the ventral surface of the embryo now overlies the ventral side of the egg (compare Figs. 24A and 24B with 24C). During this same time interval the germ band has contracted so that the embryo is restricted to the ventral surface of the egg. This contraction is also linked to proctodeal and mouthpart development. After germ band contraction the proctodeum, which appears in the nineteenth segment, comes to lie at the posterior pole with its invagination directed anteriorly (Figs. 24C and 24D). Segments 17, 18, and 19 eventually fuse to become the eighth abdominal segment. That portion contributed by the eighteenth segment grows posteriorly as a separate unit from the dorsal surface of the eighth abdominal segment. This is to become the respiratory siphon. The ninth abdominal segment forms the ventral or anal lobe. At the anterior end of the embryo there is a dramatic shift in the positioning of the mouthparts. The three gnathal segments migrate anteriorly to underlie the cephalic region (compare Figs. 24B and 24C).

The embryonic body form is completed by the process of dorsal closure. This is the gradual dorsal and medial growth of the body wall until the two walls fuse mid-dorsally to completely enclose the yolk mass. The labral segment is closed almost at the time of its formation; the cephalic segment closes between stages 36 and 46. At 46 hours a cross section of the thorax and abdomen appears hemispheric. However, by stage 56 the body wall has extended to a point three-fourths of the way to the mid-dorsal line. This dorsad growth continues, and by 66 hours closure of the posterior abdominal segments has begun; five hours later the process has begun in the anterior segments. By stage 76 dorsal closure is completed. At this time the epidermal layer is seen to infold at specific intervals which correspond to the already determined ventral intersegmental areas (Fig. 26).

HEAD SEGMENTATION

Labrum. The mouthpart rudiments are among the first structures to appear in the embryo. At the anterior embryonic margin the labrum first appears as a bilobed structure (Fig. 27). A medial invagination of the germ band projects dorsally and fuses with the dorso-lateral extensions of the germ band. The lobes remain attached to the medial walls. More posteriorly the invagination becomes progressively less pronounced; it disappears at the middle of the segment. The posterior margin of the labrum is identified by the postero-ventrally oriented stomodeal invagination. Mesodermal cells are present in the labral segment at the termination of gastrulation and do not migrate there from a post-oral position. These cells will eventually give rise to the labral musculature. At 46 hours the labral surface takes on a sculptured appearance — probably indicative of secretory activity of the epidermal cells. Ten hours later the labrum becomes laterally compressed. Its lateral walls greatly thicken and begin to buckle inward (Fig. 28). This invaginated area is initially directed laterally. However, with a gradual dorsal movement of the hypodermis and consequent dorsal-ventral flattening of the labral segment, it becomes directed ventrally. During this shift in position the cells of this invaginated region begin to send out cytoplasmic processes (Fig. 28). These processes develop into the feeding brushes, which by stage 76 are directed ventrally and will eventually fill the space between the ventral labrum and eggshell. Also at stage 76 the above movements produced an outfolding of the medial hypodermis located between the oral brushes. In cross sections of the labrum this outfolding has a tear-drop shape with many hairs projecting from the surface (Fig. 29). It has a cavity which is partially filled with cells. This structure is the palatum. At stage 96 it appears along most of the labral length, and its point of attachment with the labrum proper is highly constricted (Fig. 29). By 81 hours the dorsal cuticle of this segment becomes strikingly undulated. This entire dorsal area is the clypeus.

Antennae. At stage 31 the antennal rudiments appear as hollow buds of ectodermal tissue on each side dorso-lateral to the stomodeal invagination (Fig. 30). They arise at the transition between the labral and cephalic segments; a distinct antennal segment is not observed. These rudiments elongate slightly and become directed ventrally due to blastokinesis. Later in development the antennae emerge from the area of the head corresponding to the gena. The point of origin is over the entire length of this structure, since they are still observed at the level of the maxillae. By 71 hours an undulated cuticle has been secreted, and the antennae have retracted somewhat into the head capsule so that they appear to arise from a pit (Fig. 31). Their three-segmented nature first appears at 81 hours.

The cephalic segment possesses no appendages.

Mandibles. The first gnathal segment which gives rise to the mandibles is located just posterior to the cephalic segment. The mandibles first appear at stage 31; each member of the pair originates as a laterally directed fold from either side of the dorsal midline. Internally the distal portion of each fold possesses a cavity containing mesodermal cells. With the anterior migration and fusion of the gnathal segments to the ventral surface of the cephalic segment, the mandibles lie lateral to the anterior border of the pre-oral cavity. After blastokinesis they continue their growth as ventro-medially directed, slightly curved lobes (Fig. 32). Secretion of a thin cuticle has begun around this structure at 66 hours. By the time of hatching the mandibles possess a rather thick cuticular covering, with many hairs originating from the distal border.

Maxillae. The early ontogeny of the maxillae or derivatives of the second gnathal segment is very similar to that of the mandibles. These structures lie directly posterior and slightly lateral to the mandibles. The major difference is that at stage 71 a small finger-like projection emanates from the anterior base of each maxilla. These are the maxillary palps. The maxillae grow ventro-medially so that at the termination of development the palps may slightly overlap each other. Both maxillae and mandibles articulate broadly with the gena to permit adductory movement.

Labium. The labial segment is the third gnathal segment and lies just posterior to the maxillae. It is unpaired and possesses no appendages. The subesophageal ganglion will develop within this structure (Fig. 33).

EMBRYONIC AND EXTRAEMBRYONIC MEMBRANES

The dorsal growth of the amnion and serosa continues until 31 hours, when fusion of each membrane has been completed along the entire length of the embryo. The serosa is the thicker and more conspicuous of the two membranes. Except in the head region, the amnion usually appears as a thin, frail membrane adherent to the serosa. Under these circumstances its presence can be detected only at the lateral germ band before it joins with the serosa. These conditions persist up to approximately 66 hours of development. At this time the amnio-serosal cells in the thoracic and abdominal region aggregate to form a mid-dorsal longitudinal clump of cells extending between the first thoracic and sixth abdominal segment (Fig. 34). This structure is the dorsal organ. Its cells possess no definite shape and appear to be undergoing degeneration, as evidenced by indistinct cell membranes and pycnotic nuclei. By 71 hours the dorsal organ is completely absent. There are still remnants of the amnion and serosa in the head region, however.

Between stages 36 and 46 the vitelline membrane is present. It appears as a dense, noncellular membrane located between the endochorion and serosa (Fig. 35). This membrane is secreted by the serosal cells and attains a thickness of 1.6 μ by 56 hours.

Organogenesis

Once the mesodermal layer has formed and segmentation has begun, a series of localized cellular proliferations and an interrelated series of cellular movements take place leading to the formation of organ rudiments. Organogenesis is that portion of embryology which traces components of individual organ systems from their first appearance to the time of hatching.

ALIMENTARY CANAL

The alimentary canal is composed of a foregut, midgut, and hindgut. The developmental history of each portion will be described separately.

Foregut. Development of the foregut begins at stage **26** with the appearance of the stomodeal invagination. This ectodermal invagination occurs in the mid-dorsal region of the pregnathal segment (Fig. 36) and divides the segment into two unequal parts, the anterior being longer than the posterior. It is directed postero-ventrally for a distance of 20 μ, and its walls are composed of columnar cells (Fig. 37). Mesodermal cells in the area of the mouthparts migrate to form a layer surrounding the stomodeal invagination (Fig. 37). This enveloping layer will be of prime importance in midgut formation. The stomodeum continues to grow apically; at stage **46** it reaches its most posterior extent at the level of the first thoracic segment. At stage **46** the pre-oral area lies ventrally and is flanked on either side by the mandibles and the maxillae. With the continued ventro-medial growth of these appendages their dorsal surface becomes the floor of the pre-oral cavity. A short distance posterior from the stomodeal opening there is a dorsal evagination of the stomodeum which is the pharyngeal rudiment (Fig. 38). The posterior margin of the pharynx is U-shaped in cross section and proceeds posteriorly for 15 μ; it then changes into an oval tube. This oval portion is the esophagus. With the appearance of the pharynx and esophagus, the stomodeum may be referred to as the foregut. At its origin the esophagus is located dorsally in the body cavity; however, it gradually slopes to a more ventral position. The esophagus is composed of a columnar epithelium which is surrounded by a layer of flattened mesodermal cells (Fig. 38). These mesodermal cells will eventually develop into a layer of circular muscle. At stage **71** the cells of the pharynx and esophagus begin to secrete a chitinous intima which is present for the remainder of embryogeny.

At stage **76** the proventriculus begins to develop. It arises from a movement of the posterior foregut into the midgut and terminates in the dorsal midgut at the level of the second thoracic segment. In longitudinal section it is seen that the proventriculus is composed of three layers. The inner and outermost layers are those of the foregut, which has folded back upon itself (Fig. 39). The middle layer is called the cardia and is derived from the mesodermal layer surrounding the foregut (Fig. 40). The cells composing the walls of the proventriculus become rearranged into longitudinal rows so that in cross section the proventriculus appears stellate. The posterior tip of the proventriculus is constricted and will function as a valve in larval life.

The salivary glands, which are derivatives of the foregut, first appeared at stage 71 as an evagination of the foregut in the neck region. By stage 86 their arrangement is easily interpreted. A common duct connects to the mid-ventral foregut at the border of the proventriculus. It then bifurcates laterally so that a smaller duct projects in each lateral direction. Two pairs of salivary glands are given off in each direction. One member of each pair is directed dorso-laterally, while the other projects ventrally. In cross section each gland is composed of two or three cells and possesses a distinct lumen. The cells are irregularly shaped and contain one large nucleus and a large nucleolus. Their cytoplasm is highly vacuolated.

Hindgut. The hindgut first appears at stage 26 as an ectodermal proctodeal invagination of the nineteenth segment (Fig. 36). This invagination occurs at the posterior margin of this segment. The pole cells which were associated with this segment now lie at the tip of the invagination. The germ band is in the elongated form so that this ventral infolding is directed toward the posterior pole of the egg. As a result of germ band contraction, the proctodeal opening relocates at the posterior tip of the egg, which now corresponds to the posterior terminus of the embryo. The tubular invagination is composed of columnar cells and is directed anteriorly (Fig. 41).

At stage 31 the proctodeal invagination lies in contact with the mesodermal layer of the underlying segments. Some of these mesodermal cells migrate dorsally to form a single layer enclosing the tube (Fig. 42). As with the stomodeal invagination, these cells will be of importance in midgut development. The most anterior extent of the proctodeal invagination is the posterior border of the fifth abdominal segment.

The first portion of the hindgut is the ileum. It connects broadly with the posterior margin of the midgut and then sharply decreases in diameter, forming a funnel-like structure referred to as the pyloric ampulla. At stage 71 a characteristic looping of the hindgut occurs just posterior to this ampulla at the level of the sixth abdominal segment. It is an S-shaped bend which may lean sharply to either the left or right side but always terminates at a level more dorsal than its anterior end. Between stages 81 and 96 the hindgut undergoes histological differentiation. The cells of the pyloric ampulla are irregularly shaped, each with a single large oval to spherical nucleus containing a single large nucleolus. Chromatin granules are present at the periphery of the nucleus. Large cytoplasmic units are pinched off from the apex of these cells and released into the lumen (Fig. 43). The anterior portion of the loop is the ileum. It is composed of a layer of attenuated cells on a distinct basement membrane and is surrounded by a thin mesodermal layer (Fig. 44). The colon or second portion of the hindgut begins about halfway through the loop. Its walls are composed of cells that have formed longitudinal wedge-shaped bands (Fig. 45). This portion is surrounded by a single layer of mesodermal cells. As it continues posteriorly, its diameter decreases so that at the anal opening the lumen is almost totally obscured.

A third division, called the rectum, is usually found in larval forms but was not observed in the embryo.

At stage 46 the rudiments of the anal papillae are first seen (Fig. 41). They arise as four posteriorly directed finger-like evaginations of the posterior margin of the hindgut and are ectodermal in origin. They are composed of a thick, noncellular outer layer with an inner cellular core (Fig. 46). In cross section papillae are found in the dorso-lateral and ventro-lateral "corners" of the hindgut. They extend slightly beyond the anal opening. The papillae continue their posterior growth and at stage 76 move dorsally so that in cross section they are arranged in a semicircle (Fig. 47). The point of origin of the papillae marks the termination of the hindgut and corresponds to the anal opening.

At stage 46 the rudiments of the malpighian tubules are seen as lumenate cords of ectoderm growing posteriorly from the tip of the proctodeal invagination (Fig. 48). The tubules at this stage are composed of six cells in cross section and have a length of 20 μ. The five malpighian tubules connect to the hindgut at the dorso-lateral, mid-lateral, and ventro-lateral regions (Fig. 49). The tubules continue to grow posteriorly; by stage 71 each tubule is composed of two cells in cross section and possesses a distinct lumen. During their subsequent ontogeny the tubules project anteriorly from their origin to the level of the fourth abdominal segment, and then bend posteriorly to terminate at the eighth segment.

Midgut. The midgut takes its origin from a migration of the mesodermal cells surrounding the stomodeal and proctodeal invaginations. Early in their ontogeny both of these ectodermal invaginations became surrounded by mesodermal cells. These mesodermal cells are derived from gastrulation and are not localized as a distinct group between the stomodeal and proctodeal invaginations in the future midgut region. At this stage they do not possess a unique cytology. Between stages 41 and 46 these same cells migrate outward as a tongue from lateral walls of each invagination (Figs. 50, 51). This spreading of mesodermal cells begins before the proctodeum has reached its definitive posterior position. The migration of these four tongues of mesodermal cells (two from the proctodeal and two from the stomodeal invaginations) continues posteriorly or anteriorly dorso-lateral to the ventral nerve cord and in contact with the mesodermal rudiment (Fig. 52). By stage 46 the anterior tongues fuse with the posterior ones at the embryo's mid-region. Even when the midgut rudiment is first established in this way, its cytology is not distinctive. By stage 56 it appears cord-like in cross section. Just lateral to the midgut cells and in contact with them is a flattened layer of splanchnic mesodermal cells derived from the adjacent mesodermal rudiment (Fig. 53). The midgut cells are columnar, with their nuclei situated at approximately the same level; the cytoplasm is highly vacuolated. After fusion the two resulting midgut cords begin to grow dorsad and laterad. When the mid-ventral and mid-dorsal fusion of the growing cords takes place (stages 71 and 76 respectively), the midgut forms a tube completely enclosing the yolk material. The cells constituting the midgut walls lose their columnar shape and become somewhat flattened and/or irregular in outline. As a result of the protrusion of the developing proventriculus into the midgut, the latter comes to enclose the posterior margin of the proventriculus.

The gastric caeca first appear at stage 81 as four out-pouchings of the anterior margin of the midgut: two dorso-lateral and two ventro-lateral caeca (Fig. 54). Four additional caeca appear in the same manner at stage 86. These caeca are located dorso- and ventro-medially and remain somewhat smaller than their lateral predecessors. The caeca possess a distinct lumen which remains confluent with the midgut lumen.

At the termination of embryogeny the midgut extends from the anterior margin of the first thoracic segment to the posterior border of the fifth abdominal segment. Its epithelial cells are rectangular in shape and surrounded by a sparse layer of splanchnic mesoderm (Fig. 55).

NERVOUS SYSTEM

Brain and associated structures. The development of the brain or supraesophageal ganglion begins at stage 46 in the cephalic segment.

It arises from localized proliferations of ectoderm in the dorso-lateral and mid-lateral regions of each side of this segment (Fig. 56). The two cell masses on each side of the segment fuse so that later in development the proliferative areas are obscured. These two masses do not represent the proto-, deuto-, or tritocerebrum; rather, they fuse as a homogeneous cell mass which subsequently becomes segmented. By stage 76 the protocerebrum and deutocerebrum have become well demarcated. The protocerebrum is the largest of the three segments and occupies the posterior half of the cephalic segment (Fig. 57). The deutocerebrum is the second-largest segment and lies anterior and slightly dorsal to the protocerebrum. At stage 51 the ventral extensions of the proto- and deutocerebrum flank both sides of esophagus and fuse with the subesophageal ganglion (Fig. 58). This results in the complete enclosure of the esophagus by the cerebrum. Projecting from the medial posterior margin of the deutocerebrum is the considerably smaller tritocerebrum (Fig. 59). Each of these brain segments is composed of left and right lobes which arise independently and fuse medially. The brain extends between the pharynx and the posterior wall of the cephalic segment. By stage 81 the brain is highly compacted within this segment, and the three segments comprising it are difficult to distinguish. At stage 46 axonal processes begin to form in the central portion of the proto- and deutocerebrum; this is the beginning of neuropile development. The neuropiles in each lobe increase in size and grow medially until at stage 56 they fuse, forming the cerebral commissure (Fig. 60). The developing brain remains in direct contact with the ectoderm until stage 71, when it becomes completely separated. By stage 96 the peripheral cells of the brain differentiate to form a thin membrane loosely surrounding each lobe of the brain (Fig. 61). This is the neurilemma. The cells constituting this membrane possess long thin cytoplasmic processes; their nuclei are spindle-shaped with a peripheral chromatin layer.

Brain development, along with that of the other parts of the nervous system, involves numerous mitoses. Individual chromosomes are not observed in the various stages of these dividing neural cells. Instead, mitotic and newly separated cells possess nuclei containing a single mass of intensely staining chromatin (Fig. 62). Mitoses are still observable at the time of hatching.

Stomodeal nervous system. The stomodeal nervous system consists of sensory and motor neurons connected to a ganglion derived from the stomodeum (Snodgrass, 1935). In the embryo of *Aedes vexans* this ganglion is the frontal ganglion. It first appears at stage 66 as cells budding off from the anterior surface of the pharynx. By stage 76 the ganglion is oval-shaped and has a central neuropile (Fig. 63). Posteriorly it gives off a cord of cells which is the recurrent nerve. This continues posteriorly along the dorsal surface of the esophagus. At the posterior level of the circumesophageal nerve it begins to enlarge to form the hypocerebral ganglion (Fig. 64), which then bifurcates into two nerves that innervate the esophagus. At stage 76 a lateral nerve extends outward from each side of the frontal ganglion and innervates the antero-medial margin of the deutocerebrum. At stage 86 a nerve appears from the ventral surface of the frontal ganglion and attaches to the hypodermis. Connected with the retrocerebral complex are two pairs of glands, the corpora allata and corpora cardiaca. These glands first appear at stage 66 as a pair of small cellular masses budding off from each lateral wall in the mid-region of the cephalic segment (Fig. 65). Each pair of developing glands then becomes detached from its ectodermal source, and the larger dorsal gland fuses with the postero-lateral margin of the protocerebrum. This gland is the corpus cardiacum and soon establishes a neuronal connection with the protocerebrum (Fig. 66). The corpus cardiacum is a bulbous structure, usually having a single layer of cells surrounding the central neuropile. The cells look much like other neural cells but contain more cytoplasm. The smaller cellular mass is the corpus allatum and is usually composed of four cells (Fig. 65). After its detachment from the wall this mass fuses to the ventral surface of the corpus cardiacum.

The development of the subesophageal ganglion begins between stages 36 and 46 in the ventral surface of the mandibular, maxillary, and labial segments. This ganglion is usually viewed as the anteromost portion of the ventral nerve cord. Within each of these segments neural cells proliferate from the ectoderm on either side of the midline. This results in a pair of spherical cellular masses in each segment (Fig. 67). Ganglionic cells have spherical nuclei with a peripheral layer of granular chromatin and very little cytoplasm; they are indistinguishable from cells of the brain. The ganglia increase in size and at stage 51 begin to separate from the ectoderm. At this same stage a neuropile begins to differentiate at the dorsal surface of each ganglionic lobe. The adjacent neuropiles fuse medially to form an intraganglionic connective (Fig. 68). At stage 56 the circumesophageal nerve forms as a result of the fusion of the tritocerebral lobes with the first ventral ganglion. The neuropiles of each of these structures join to form a ring completely surrounding the esophagus (Fig. 58). From the time this structure first differentiates until the end of embryogeny, the axonal processes are always surrounded by ganglionic cells. At no time does the circumesophageal nerve appear as a ring composed of only axonal processes.

Axonal processes grow out perpendicularly from the neuropile of each ganglionic lobe and fuse with similar outgrowths of adjacent lobes; these are the interganglionic connectives (Fig. 69). The ganglia of the mandibular and maxillary segments migrate posteriorly and at stage 71 are located entirely within the labial segment. The circumesophageal nerve is now found at the anterior margin of this segment (Fig. 58).

At stage 81 the three segments of the subesophageal ganglion have fused, and their trisegmented origin is no longer detectable (Fig. 54). The entire subesophageal ganglion becomes enclosed within a loose sheath derived from the peripheral ganglionic cells. This sheath is the neurilemma. Ganglionic cells of the anterior border of the deutocerebrum connect with the antennae at stage 51. Fiber tracts between these structures were not observed at stage 96.

Ventral nerve cord. Development of the ventral nerve cord in the thoracic and abdominal segments begins at stage 26 with the production of neuroblasts. After the mesodermal layer has formed, neuroblasts migrate from the ectoderm and come to lie between the ectodermal and mesodermal layer (Fig. 70). These cells do not arise from mitotic divisions of the surface ectoderm. Also at this stage a neural groove begins to extend longitudinally along the middle of the ectodermal layer (Fig. 70). Once separated from the ectoderm, the neuroblasts begin to divide. These cells are somewhat irregular in shape, with large spherical nuclei and diffuse chromatin. By stage 46 divisions have progressed to the point where obvious ganglionic lobes are present on each side of the neural groove. An antero-posterior gradient seems to exist in the development of the nerve cord, with the anterior portion more advanced in development than the posterior one. At stage 51 a neuropile begins to differentiate in the dorsal portions of the anterior ganglionic lobes. In longitudinal section the anterior and posterior borders of the neuropile send off visible ventral extensions (Fig. 71). As development proceeds, the neuropile increases until it is a single large mass at the time of hatching. Interganglionic commissures and connectives are present by stage 66. The anterior connectives of the first thoracic segment are attached to those of the subesophageal ganglion (Fig. 58). In all there are three thoracic and eight abdominal ganglia. Throughout the embryonic period all eleven ganglia remain unfused and attached to each other by the interganglionic connectives (Fig. 72).

Larval Eyes. At stage 76 on each side a mid-lateral spherical area of the epidermis at the level of the pharynx begins to differentiate into a larval eye. The nuclei of this epidermal layer of each eye rudiment migrate to the base of each cell so that the outer portion of each eye is clear cytoplasm (Fig. 73). Five hours later rust-colored pigment granules appear in the cytoplasm (Fig. 74). By stage 86 each vertically situated eye has a basal layer of ten nuclei at its greatest length. Above the nuclei and extending to the cuticle is a dense layer of pigment granules. A few flattened nuclei are occasionally noted at the surface beneath the cuticle. Optic tracts were not observed.

OTHER ECTODERMAL DERIVATIVES

The enclosing ectodermal layer gives rise to a number of structures in addition to those already described. Among these are the cuticular covering, cuticular specializations such as the hatching spine, the tracheal system, certain hemocytes, and the imaginal discs.

Cuticle and cuticular specializations. At stage 56 the epidermal cells in the cephalic region, antennae, and in the seventh and eighth abdominal segments begin to secrete a thin cuticular layer. These cells are somewhat irregular in appearance but become cuboidal to columnar by stage 66. At stage 66 a cuticular layer is secreted by the mandibles and maxillae; the mandibles in particular possess a highly thickened cuticle at hatching. During the secretory process these epidermal cells often have a vacuolated cytoplasm. Between stages 71 and 81 the dorsal cuticle of the labrum, which is the clypeus, becomes highly undulated and thickened (Fig. 75). The epidermal cells are in contact with the cuticular layer. These cells are ameboid in shape, with a vacuolated cytoplasm and spherical to oval nuclei (Fig. 75). In the cephalic segment the epidermal cells are rectangular in shape and possess oval nuclei. The epidermal cells in the thoracic and abdominal segments vary in shape according to their location. The cells of the lateral walls are large and elongate with elongate nuclei; they lie parallel to the surface. Those of the dorsal and ventral surfaces are highly attenuated. Between stages 91 and 96 the thin cuticular covering of the thoracic and abdominal surfaces is secreted. Cuticular specializations such as the labral brushes and mandibular hairs have already been discussed.

At stage 51 a spherical area of the dorsal epidermis at the posterior level of the maxillae begins to differentiate. This eventually will give rise to the hatching spine or egg breaker. First it sinks below the level of the epidermis and its cells begin to change their shape. In a longitudinal section through the diameter of this disc the cells in the central region are highly columnar, whereas peripherally they grade into a flattened type (Figs. 59, 76). At stage 66 this structure begins to secrete a thick cuticle. The central cells secrete more cuticle than the peripheral ones; this difference is responsible for the development of the spine. At the time of hatching the spine is located dorsal to the pharynx. In longitudinal section it appears triangular, with its two walls composed of dense cuticle and a cellular base (Fig. 77). The apex of the spine is a pointed mass of cuticle.

Oenocytes. Development of the oenocytes takes place at stage 56. They appear in somewhat elongate bundles of three or four cells differentiated from the ectoderm in the anterior, mid-lateral walls of the abdominal segments (Fig. 78). These cells are located beneath the mid-lateral epidermis and are surrounded medially by a thin dorso-ventrally oriented muscle (Fig. 78). The oenocytes are spherical and considerably larger than the surrounding cells. They have a single large oval nucleus with a peripheral ring of granular chromatin and a single nucleolus. The cytoplasm stains a pale gray and is slightly vacuolated. The oenocytes remain in this position throughout embryogeny.

Imaginal cells. Occasionally small clusters of cells appear in relation with the walls of the thoracic and abdominal segments. The cells appear compressed and have intensely staining nuclei with a scanty layer of surrounding cytoplasm. These clusters are probably the imaginal discs. A detailed description of their location and potentialities must be deferred until a thorough analysis of the larval form is completed.

Tracheal system. The tracheal rudiments of the respiratory siphon are the first to appear at stage 46. They arise as a ventrally directed ectodermal invagination on either side of the dorsal mid-line just anterior to the point where the respiratory siphon extends from the eighth abdominal segment (Fig. 79). Each rudiment is lumenate and opens to the dorsal surface. Later in development these surface openings close off, and the tracheal trunks connect to the longitudinal trachea of the seventh abdominal segment. The trunks grow posteriorly and at stage 71 form an S-shaped bend (Fig. 80) so that in a cross section of the siphon each trachea appears three times. At stage 76 each tracheal trunk is surrounded by a mass of ectodermal cells which later differentiate into tracheoles (Fig. 81). A connection between these tracheoles and the tracheal trunk was not observed. By stage 91 the two ends of the tracheal trunks have fused and open to the surface at the posterior tip of the respiratory siphon.

The tracheal rudiments of the thoracic and abdominal segments appear at stage 71 in conjunction with the invaginations of the ectodermal layer which demarcate the segments (Fig. 26). This infolding occurs along the dorsal surface and terminates at the mid-lateral surface of each segment; the invagination is not present at the dorsal midline. In longitudinal section, at the apex of the infolding, there are two or three very small ectodermal cells (Fig. 26). These cells form a semicircular band around the dorsal half of the midgut in the thoracic and first seven abdominal segments. Between stages 86 and 91 the cells differentiate into tracheal tubes (Fig. 82).

Also, a longitudinal tracheal trunk develops on each side of the dorsal midline and connects with the ventral-laterally extending trachea of each segment (Fig. 83). At the anterior margin of abdominal segment 8 each longitudinal trunk becomes fused with one of the trunks in the respiratory siphon. Anteriorly the trunks enter through the neck and there undergo much branching. Trachea are observed between and around the lobes of the brain and at the base of the larval eyes. The origin of these trachea is uncertain.

MUSCULAR SYSTEM

Since the mosquito larva possesses a large number of muscles, many of which are minute, only the larger and more important ones will be described by their specific names. The nomenclature used is that of Christophers (1960).

The development of the embryonic musculature begins in the cephalic area. At stage 31 mesodermal cells are located at the posterior level of the labral segment and are in contact with the dorsal ectoderm. By stage 46 these cells become disassociated and are found in the yolk mass (Fig. 84). They will develop into the labral musculature. At stage 56 some of these cells aggregate to form a dorso-ventrally oriented cord on each side of the dorsal midline (Fig. 85). Shortly thereafter another cord forms just lateral to each of these. Both pairs of cords soon differentiate into the median and lateral retractors of the flabella (Fig. 86). These are the two largest muscles in the embryo. Each pair of retractors (i.e., one medial and one lateral muscle) inserts at a common point on the ventro-lateral hypodermis of the posterior labrum (Fig. 86). By the end of embryogeny this point of attachment becomes a thick apodeme. The lateral retractors take their origin posterior to their insertion in the dorso-lateral epidermis. The medial retractors have their origin in the mid-dorsal epidermis posterior to the origin of the lateral muscles. This point of origin also coincides with the anterior peripheral cuticle of the hatching spine.

The muscles of the appendicular structures are derived from mesodermal cells located within the cavity of these structures. In the case of the mouthparts, mesodermal cells move into these rudiments before the mouthparts migrate anteriorly. At stage 76 the antennal muscles develop between the base of the antennae and an area below the frontal ganglion. Also at this stage the major muscles associated with the mandibles and maxillae appear. The adductors of the mandible are muscle bands which insert on the postero-medial edge of the mandibular base and take their origin at the lateral wall just ventral to the developing eye (occipital region) (Fig. 87). The abductors of the mandible have the same point of origin as the adductors but insert on the lateral edge of the mandibular base (Fig. 88). The depressors of the maxillae are muscles extending between the occipital region and the dorsal portion of the maxillary base (Fig. 89). The opposing retractors of the maxillae are thin muscles connecting between the lower lateral cephalic wall and the base of the maxilla (Fig. 89).

At stage 81 the dorsal retractors of the pharynx appear. These are a thin pair of muscles inserting at the postero-dorsal margin of the pharynx and originating in the dorsal cuticle just posterior to the hatching spine. These are the decussating muscles of the pharynx. One extends from each lateral edge of the hatching spine, decussates ventrally, and connects to the dorsal surface of the pharynx. The circular muscles of the foregut are derived from the mesoderm surrounding the stomodeum shortly after its invagination (Fig. 37).

A complex of muscles appears in the neck region at stage 81. These are primarily small muscles extending from the neck into the thorax; a few reach into the head capsule.

The musculature of the thoracic and first seven abdominal segments is derived from the lateral mesodermal rudiments. At stage 46 these rudiments appear as a plate two or three cells thick, located dorso-laterally to each ganglionic lobe. As the anterior and posterior tongues of midgut cells migrate, they come into contact with this mesodermal rudiment (Figs. 52, 53). The inner layer of cells of this rudiment split off to become the splanchnic mesoderm; this layer becomes adherent to the outer surface of the midgut (Fig. 53). The splanchnic mesoderm grows dorsally and ventrally with the midgut rudiment so that when the latter becomes a closed tube it is enclosed by a sparse layer of splanchnic mesoderm (Fig. 55). At no time during embryogeny, however, do these mesodermal cells differentiate into muscle or any other identifiable structure. The somatic mesoderm forms from those cells of the mesodermal rudiment remaining after the splanchnic mesoderm splits off; they will differentiate into body wall musculature. The somatic mesoderm grows dorsally, keeping pace with the growth of the body wall. Between stages 71 and 76 the layers of somatic mesoderm meet mid-dorsally. At stage 71 the mesodermal cells below the gut begin to differentiate into the ventral longitudinal series of muscles. The muscles of adjacent segments connect to the cuticular invaginations demarcating intersegmental regions. This same arrangement exists for the dorsal longitudinal series of muscles which appear by stage 81. Each of these series consists of several bundles, which may be arranged in three directions. These arrangements are similar for the thoracic and abdominal segments. The more dorsal muscles in each series are directed antero-posteriorly, while the deeper ones are oriented obliquely in opposite directions (Figs. 90, 91). At stage 76 a dorso-ventral series of muscles begins to differentiate in the mid-lateral wall of the thoracic and first seven abdominal segments (Fig. 92).

The layer of circular muscle surrounding certain portions of the hindgut is derived from the mesodermal cells that were present at the time of the proctodeal invagination (Figs. 41, 42). The mesodermal cells associated with embryonic segments 17, 18, and 19 give rise to the musculature of the eighth abdominal segment with its associated respiratory siphon and anal lobe. The development of these muscles begins at stage 71.

CIRCULATORY SYSTEM

The heart develops at stage 76 when the two layers of somatic mesoderm meet mid-dorsally. In cross section the heart is composed of two adjacent ameboid-shaped cells, one donated by each of these layers. By stage 86 these cells have become crescent-shaped with a central nucleus. A small lumen is present between the two cells. By stage 91 the cells become thinned out and the nucleus projects into the lumen (Fig. 93). The heart lies dorsal to the gut and is flanked on either side by the dorsal longitudinal muscles. Anteriorly it extends through a groove between the two lobes of the brain and terminates shortly before reaching the circumesophageal nerve. Posteriorly the heart ends at the base of the respiratory siphon and reaches its greatest diameter at the level of the hindgut (Fig. 93). In the larval form the portion of the heart anterior to the first abdominal segment is called the aorta. This segment is distinguishable from the heart in that the former lacks ostia, allary muscles, and pericardial cells. Such a distinction is not observable in the embryo.

FAT BODY

The fat body first appears at stage 71 and is derived from the outer cells of the somatic mesoderm. Fat body cells are present in the thoracic and abdominal segments and are positioned between the epidermis and somatic mesoderm in the lateral and dorsal regions of these segments (Fig. 94). The nuclei of these cells are ameboid, with granular chromatin evenly distributed throughout; the cytoplasm is reticular (Fig. 94). This organ remains unchanged for the remainder of the embryonic period.

GONADS AND POLE CELLS

The gonads are the only portion of the reproductive system to appear during embryogeny.

Before completion of germ band contraction the proctodeal invagination is directed toward the posterior pole of the egg, with its tip at the level of the sixth through seventh abdominal segments (Fig. 95). At this time the pole cells have divided into three separate groups. The two lateral groups located at the tip of the proctodeum (plus the mesodermal sheath which develops later) will give rise to the gonads. They migrate laterally and come to lie against the mid-lateral walls on each side of abdominal segments 6 and 7 (Fig. 96). Since the cells of these two groups will form part of the gonads, they may be referred to as primordial germ cells (abbreviated PGC's). Medially the PGC's are in contact with the posterior midgut and laterally with the mesodermal rudiment (Fig. 96). There is a slight anterior migration so that by stage 61 these cells are localized within segment 6. By stage 56 each group of cells has formed a compact bundle which tapers anteriorly and is enclosed by a layer of attenuated mesodermal cells. These cells are derived from the medial cells of the mesodermal rudiment. At the posterior tip of each gonad the mesodermal cells form a filament which attaches to the postero-ventral border of the seventh abdominal segment. An anterior filament was not observed. The cytology of the primordial germ cells does not change significantly during development; usually there is only a decrease in the amount of cytoplasm. At stage 66 some of these cells appear to be entering prophase, and by stage 81 the number of cells constituting each gonad has increased slightly. At stage 96 each gonad consists of ten PGC's and is slightly compressed laterally; the ends are tapered. Measurements of the diameters of the gonads in stages 81-96 showed that differences may exist within the same stage. However, since so few specimens were available, the significance of these differences could not be determined.

The third group of pole cells, unlike the other two, plays no role in the development of the gonads. By stage 46 these four cells are found in the postero-dorsal portion of the anal lobe, just dorsal to the anal papillae (Fig. 46). They apparently migrate to this position during germ band contraction. Unlike the development of the gonads, where the number of initial pole cells is somewhat variable, the number found in this location is always four. These cells are still present in this position at the time of hatching.

The time of occurrence of embryonic processes and appearance of major embryonic structures, along with their germ layer derivation, are summarized in Table 1.

TABLE 1

TIME OF OCCURRENCES OF EMBRYONIC PROCESSES AND APPEARANCE OF MAJOR
EMBRYONIC STRUCTURES ALONG WITH THEIR GERM LAYER DERIVATION

Stage	Event / Structure	Germ layer
1	Polar body formation	—
2	Fertilization	—
2-3	First cleavage	—
5	Pole cells, primary yolk	—
6	Peripheral migration of cleavage nuclei	—
13	Secondary yolk cells	—
15	Germ band formation	—
18½	Amnion, serosa, germ band elongation, gastrulation	—
21	Segmentation, labrum	—
26	Proctodeal and stomodeal invaginations, nerve cord	—
31	Antennae, mandibles, maxillae	—
36-46	Germ band contraction, blastokinesis, vitelline membrane, labrum	—
46	Midgut	mesoderm
	Pharynx	ectoderm
	Brain	ectoderm
	Malpighian tubules	ectoderm
	Anal papillae	ectoderm
	Trachea (respiratory siphon)	ectoderm
51	Gonads	pole cells
	Hatching spine	ectoderm
56	Cuticle (cephalic) labral musculature	mesoderm
	Oenocytes	ectoderm
66	Frontal ganglion dorsal organ	ectoderm
	Corpora cardiaca and allata	ectoderm
71	Trachea (body), salivary glands	ectoderm
	Fat body	mesoderm
76	Proventriculus, larval eye, dorsal closure	ectoderm
	Heart	somatic mesoderm
81	Gastric caecae	mesoderm
	Thoracic and abdominal muscles	somatic mesoderm
91	Cuticle (body)	ectoderm

DISCUSSION

The histological technique employed in this study seemed well suited for embryological studies on insects. The use of methacrylates as an embedding compound proved to be far superior to paraffin. Since its hardness approximates that of the endochorion, the tearing and distortion of tissue adjacent to the eggshell was eliminated. Sections can be cut at thicknesses of less than 0.05 μ. This is particularly important in embryological studies, since it permits viewing of every cell layer within each animal. This greatly reduces the number of specimens that need to be prepared within a specific stage. Harber (1969) employed a similar technique with her work on *Culiseta inornata,* but she seems to have had difficulty in obtaining serial sections. In the present study ribbons of up to sixty sections were obtained routinely. This success was probably due to the use of a diamond knife, which provided a highly uniform cutting edge. Also, the advance mechanism in the ultramicrotome is far more precise than in a modified standard microtome.

In general, the structure of the egg of *Aedes vexans* at oviposition closely resembles the description of the eggs of other mosquitoes. One important difference found was the initial absence of the vitelline membrane. This structure appears between stages 36 and 46 and often reaches a thickness equivalent to that of the endochorion. It is secreted by the serosa. With the exception of Telford (1957) and the present study, all authors working on the embryos of mosquitoes and other dipterans reported the initial presence of a vitelline membrane. Telford (1957) noted that in pasture *Aedes* this membrane appeared at approximately twenty-four hours after oviposition. He believed it to be derived from the periplasm but offered no evidence to support this view. The vitelline membrane is impermeable to water and serves to prevent the egg and its contents from desiccation. The delay of its development in the genus *Aedes* would seem to be of no particular advantage. However, since these eggs are always subjected to long periods of desiccation, it is imperative that this membrane be highly impermeable. The thin vitelline membrane of most insects is secreted by the egg itself (Korschelt and Heider, 1899). Its secretion by the serosa produces a much thicker membrane and therefore a greater barrier to water loss than if it were secreted by the egg. Hence, this group has developed a special method for the production of this membrane.

The egg of *Aedes vexans* is also unusual in that it lacks a periplasmic layer. The egg contents, with the exception of the size distribution of yolk spheres, is homogeneous throughout. Yolk spheres were frequently observed at all areas of the egg surface in direct contact with the endochorion. Korschelt and Heider (1899) stated that the periplasmic layer in insects is usually thin, while Counce (1961) described it as thin in hemimetabolous insects and well developed in holometabolous ones. Much of the literature seems to contradict Counce's statement. Various authors working on mosquitoes have described this as being existent but usually thin (Ivanova-Kazas, 1947; Larsen, 1952; Telford, 1957). Of the major works done on other dipterans, only Davis (1967) and Auten (1934) reported any significant periplasm. Counce (1961) believed that in early developmental stages the cortex or periplasm is deleterious to nuclei entering it and thus is a deciding factor in which meiotic product will become the female pronucleus. At the time of blastoderm formation the periplasm seems to be involved in the primary differentiation of the cleavage nuclei (Bhaskaran *et al.*, 1970). According to Counce (1961), maintenance of the integrity of the cortex or periplasm is essential for normal development. A disruption produces abnormalities in nuclear distribution during blastoderm formation; nuclei fail to enter the cortex, and the failure of nuclei from adjacent areas to fill in these gaps leads to localized cortical breakdown. In view of the layer's apparent importance in embryonic development, additional studies on its nature seem called for. This information might help explain why the periplasm is so large in some animals and apparently absent in others.

The newly oviposited egg of *Aedes vexans* lacks a distinct egg nucleus. However, this structure usually becomes visible at stage ½ or 1. This confirms a similar observation by Nicholson (1921) on *Anopheles maculipennis*. He found that, at the time of oviposition, the chromatin of the egg nucleus is dispersed throughout most of the egg. Within a short time the chromatin becomes localized and the female pronucleus is formed. Prior to fertilization the pronucleus in *Aedes vexans* usually migrates to the surface and undergoes meiosis I to form the first polar body. Only the first polar body undergoes meiosis II so that two polar bodies are formed. This observation is in agreement with that of Davis (1967) on *Culex fatigans*. Uncommonly, a single polar body is formed in *Lucilia sericata* (Fish, 1947a) and *Simulium* (Gambrell, 1933), but generally three is the usual number as in *Drosophila* (Sonnenblick, 1950).

In *Aedes vexans* primary and secondary yolk cells were distinguished. The primary yolk cells are those cleavage nuclei which fail to migrate to the surface at blastoderm formation. They remain in the yolk matrix and their chromatin becomes highly compacted. In the Culicidae these cells have been observed by Davis (1967), Harber and Mutchmor (1970), Larsen (1952), and Idris (1960). Ivanova-Kazas (1947) described yolk cells in the embryo of *Anopheles maculipennis* but was uncertain whether they were primary or secondary. Primary yolk cells are common in almost all of the other dipteran embryos studied. In *Lucilia sericata*, Fish (1947a) and Davis (1967) reported that all the cleavage nuclei move toward the surface but that a few fail to invade the periplasm. Davis referred to these cells as primary vitellophages, whereas Fish stated that these were not true primary yolk cells since none of the cleavage nuclei remained *in situ* in the yolk. DuBois (1932) and Butt (1934) found that in *Sciara* all of the nuclei of the somatic line reach the superficial layer of the egg; later, several nuclei separate themselves from the blastodermic layer and penetrate into the yolk. These are the vitellophages. Based on their origin they most closely resemble secondary yolk cells. In *Aedes vexans* the secondary yolk cells are those blastodermal cells (before cell boundary formation) which migrate back into the yolk matrix. Such cells may migrate from any area of the blastoderm; there is no polarity in this cellular migration. Similar observations were made on *Culex pipiens* (Idris, 1960) and *Culex fatigans* (Davis, 1967). The latter author reported that the formation of secondary yolk cells was confined to the anterior and posterior poles. Anderson (1962b) found that the secondary yolk cells in *Dacus tryoni* arise from the posterior pole. Tertiary yolk cells have not been observed in *Aedes vexans* and other Culicidae but are described for other dipterans. In *Lucilia sericata*,

Davis (1967) reported that tertiary yolk cells were derived from the midgut rudiments during gastrulation, while in *Drosophila melanogaster* Rabinowitz (1941) found that these yolk cells had their origin in the nuclei of pole cells which had migrated into the yolk between the blastoderm nuclei. The mitotic behavior of the yolk cells is usually distinct from that of cleavage nuclei. In *Aedes vexans* these cells never divide in synchrony with the cleavage nuclei, and throughout their developmental history they were never observed in mitosis. This latter observation leads to the conclusion that these cells divide by amitosis. Auten (1934) stated that there may be amitotic division of yolk cells in *Phormia regina;* Rabinowitz (1941) found that in *Drosophila,* at the pre-blastoderm stage, they divide by mitosis but undergo amitosis during the three blasteme mitoses. Johannsen and Butt (1941) stated that yolk cells may undergo amitosis during senescence.

In *Aedes vexans* the pole cells become differentiated at stage 6 from cleavage nuclei entering the polar region. They are never part of the blastoderm and are never derived as products of mitosis from other pole cells. The pole cells are large, irregularly shaped cells containing polar granules and yolk spheres in the cytoplasm. The polar granules within the pole cells were often seen to coalesce, forming a few larger granules (about eleven of them), and these divide for the first time just prior to hatching. Ivanova-Kazas (1947) reported that pole cells first lie at the same level as the blastoderm and later push from the blastoderm while the latter closes below them. She also found that, at the time the internal pole cells separate from the blastoderm, they divide until a group of 20-30 is reached. Harber and Mutchmor (1970) counted six pole cells and found them to be dividing asynchronously with relation to each other and to the cleavage nuclei. Davis (1967) found 12-16 pole cells in *Culex fatigans.* Idris (1960) saw twelve for *Culex pipiens;* Davis indicated that these cells undergo mitotic divisions. The results of Guichard (1971) on *Culex pipiens* are in agreement with the present study in that, once the pole cells differentiate at the posterior pole, they do not divide before their migration. Generally fewer pole cells are produced within the Nematocera than in Cyclorrhapha (higher Diptera). In the latter suborder an average of 55 pole cells were counted in *Drosophila* (Sonnenblick, 1950) and 32 in *Dacus tryoni* (Anderson, 1962b). West *et al.* (1968), however, found only about eight pole cells in *Musca domestica.* Most authors agree that pole cells are more voluminous and possess much larger nuclei than somatic cells. The presence of polar granules in the cytoplasm is also characteristic. The fusion of the polar granules reported here supports the findings of Counce (1963), Davis (1967), and

Mahowald (1962). The function of these RNA-rich granules is unknown, although Bantock (1961) believed that in *Mayetiola destructor*
they prevent chromosomal elimination from taking place. The studies
of Weismann (1863) and Hasper (1911) reported the presence of
yolk spheres within the pole cells. This same phenomenon has been
observed in *Drosophila gibberosa* (Counce, 1963).

Among the Diptera the scheme for entry of the pole cells into the
embryo and migration to their definitive position is similar to the description in *Aedes vexans*. The minor points of variance do not warrant
discussion here; the fate of these cells is important, however. In all
mosquito embryos studied to date, with the exception of *Aedes vexans*,
all of the pole cells become primordial germ cells and locate in the
larval gonad. *Aedes vexans* is unique among the Diptera in that, in
addition to two groups of pole cells going into the formation of the
gonads, a third group of four cells locates in the dorsal region of the
anal lobe. No other embryological study in the Diptera to date has
reported pole cells present in this location. In *Drosophila, Dacus tryoni*,
and *Lucilia cuprina*, however, pole cells have been shown to take part
in gut development (Poulson and Waterhouse, 1960; Anderson, 1962b).
Sonnenblick (1950) observed that there are two methods by which the
pole cells enter into the embryo. The first involves an interblastodermal
migration of a variable number of free, detached pole cells from the
polar cap into the central yolk mass; the cells degenerate and disappear. Of the pole cells that enter through the posterior invagination,
some are involved in the formation of the gonads, while others become
trapped in the gut wall or lost in the yolk (Huettner, 1940). Poulson
(1947) stated that in *Drosophila* some of the pole cells form part of
the mid-section of the midgut. Poulson and Waterhouse (1960) proved
this experimentally in *Drosophila melanogaster* and *Lucilia cuprina*.
At the pole cell stage they irradiated the polar region of these eggs
with ultraviolet light. This treatment resulted in a reduction in the
number of cuprophilic cells in the middle midgut, as well as a reduction in the gonadal size. Anderson (1962b) observed that in *Dacus
tryoni* the pole cells are divided into two histologically distinct groups
based on nuclear size and staining properties. One group will develop
into the gonads, while the other divides and forms the ventral wall of
the distal end of the proctodeum. The fact that the pole cells in the
anal lobe of *Aedes vexans* have not degenerated by the termination
of embryogeny would seem to indicate that they will somehow become
involved in larval development. This is of interest, for it implies yet
another function for these cells that were thought to form either
gonads or midgut. At present it is difficult to hypothesize their fate,

since larval studies on the imaginal discs of the anal segment show that they develop from the hypoderm rather than from the dorsal ectoderm (Christophers, 1960). The only certain method for determining the fate of these cells would be to follow their development through the larval and pupal stages.

In addition to contributing to the development of specific structures, the pole cells have been implicated in exerting a physiological effect on other cells. Agrell (1963) suggested that in *Calliphora erythrocephala* the pole cells may induce a mitotic gradient in the embryo during the cell cycles directly after the pole cells are formed. The gonads of all Diptera are encased in a thin mesodermal sheath. Poulson and Waterhouse (1960) showed that this gonadal sheath arises in response to the inductive action of the germ cells that aggregate in the lateral mesoderm.

In most nematocerous and cyclorrhaphous insects definitive segmentation is preceded by a series of transitory furrows. In *Aedes vexans* such transitory furrows do not exist, and the only segmentation observed is the definitive segmentation which first appears in the elongate germ band. Each body segment is distinguishable as a slight indentation of the surface of the germ band and a coincidentally pronounced constriction of the mesodermal layer. The developing mouthparts have more obvious intersegmental areas. In *Culex pipiens* Guichard (1971) described the appearance of the cephalic furrow in the anterior third of the newly formed germ band. Shortly thereafter seven transverse furrows form in the direction of the hindgut. As the germ band begins its elongation process, these furrows disappear in a posterior to anterior direction; the cephalic furrow persists. Guichard (1971) did not believe that the initial furrows delimit the macrosegments and that only the cephalic furrow has a relationship with metamerism. Larsen (1952) observed these furrows in *Culiseta inornata* and stated that they probably indicate the boundary of the future body segments. According to Rosay (1959), the proliferation of cells creates more material than the embryonic rudiments can accommodate; this causes the formation of furrows which resemble superficial segmentation. She also found that the number of metameres and the shape of the germ band are variable during this formative stage. These transitory furrows were also observed in mosquito embryos by Ivanova-Kazas (1947), Idris (1960), and Davis (1967), and they are present in higher Diptera such as *Dacus tryoni* (Anderson, 1962b), *Musca* (Escherich, 1901), and *Calliphora* (Noack, 1901). In *Drosophila* the cephalic furrow has been observed by Ede (1956), Ede and Counce (1956), Poulson (1937), and Sonnenblick (1950). In this ani-

mal, however, the other body furrows are confined to the extraembry-
onic ectoderm in front of the forward-pushing posterior end of the
germ band (Anderson, 1962b). The early appearance and position of
the cephalic furrow appears to be constant in all the species studied.
There is, however, controversy over which segments this furrow separates. According to Anderson (1962b), in *Dacus tryoni* it lies at the
boundary between the future maxillary and labial segments of the
head. Poulson (1937) regarded the furrow as lying between the head
and trunk, whereas Idris (1960) and Davis (1967) believed that it
separates the mandibles and maxillae. It has been suggested by Sonnenblick (1950) that this furrow may act as a stabilizing factor that
offsets the stresses resulting from the diverse surface changes and cellular translocations taking place in various areas of the embryo. Some
evidence to support this theory has come from Ede (1956), who studied
the effects of the sex-linked lethal Lff11 on the embryogeny of *Drosophila melanogaster*. He found that one of the effects of this lethal is
the exaggeration and persistence of the cephalic furrow and that this
was one of the factors leading to extreme disorganization of the late
embryo. The fact that these furrows do not exist in the embryo of
Aedes vexans could undermine certain authors' belief that they serve
an important morphogenetic function. Since the histological technique
employed in the study produced essentially no tissue compression, it
is unlikely that such furrows would have gone unnoticed.

The process of embryonic segmentation and mouthpart development
as described for *Aedes vexans* is in close agreement with Guichard's
(1971) description for *Culex pipiens* and varies only slightly with other
descriptions of mosquito embryology. In *Aedes vexans* the head was
found to consist of six segments. The labrum, antennae, and cephalon
are the pregnathal segments, while the mandibles, maxillae, and labium
are the gnathal segments. The three thoracic segments remain distinct
throughout embryogeny; while the last three of the ten abdominal segments fuse to become the eighth segment. One point of divergence in
the segmental pattern of *Aedes vexans* from certain other animals is
the absence of an intercalary segment. Ivanova-Kazas (1947) distinguished an intercalary segment in the embryonic head of *Anopheles
maculipennis;* Idris (1960) also distinguished this segment immediately
posterior to the stomodeum in *Culex pipiens*. The former author found
this segment to be very small and stated that, following development
of the stomodeum, the intercalary segment sinks into the stomodeum
and probably takes part in its construction. The intercalary segment
is also referred to as the second antennal segment and is supposed to
correspond to the second antennal segment of the insectan ancestor

(Snodgrass, 1960). DuPorte (1963), however, supported the view of a protoonchophoran origin of insects; with this theory there is no justification in assigning a second antennal segment, because nowhere in the Onchophora-Myriapoda-Insecta stock is there any evidence of a second antenna. The results of this study support DuPorte's view, although some phylogeneticists might take the presence of a tritocerebrum as evidence for the existence of an intercalary segment.

Butt (1957) discussed head segmentation and stated that the labral lobes in insects form from material that moves around the stomodeum where they fuse. Our observations on *Aedes vexans* clearly do not support this view, since the labrum always arises and remains as the anteromost segment. Butt also stated that each segment has, early in its development, a pair of coelomic sacs. This is also in conflict with our observation that coelomic sacs do not exist at any time in *Aedes vexans*.

The presumptive mesoderm in *Aedes vexans* is a mid-longitudinal band at the germ band stage. At gastrulation these mesodermal cells migrate ventrally and form a solid cord, while the ectodermal cells move medially and fuse. This cord then flattens and migrates laterally to form a layer one or two cells thick. A gastrular groove does not appear. These observations are in general agreement with Idris (1960), Davis (1967), and Guichard (1971), but they differ in that these authors noted the presence of a gastrular groove. Larsen (1952) reported that in *Culiseta inornata* the groove first appears in the mid-anterior region. The fact that the formation and closure of a gastrular furrow was not observed in *Aedes vexans* could have resulted from the occurrence of these processes in between the prepared stages. Gastrulation lasts approximately one hour (Guichard, 1971); at the time gastrulation occurred in *Aedes vexans*, two-hour intervals were used between the stages. An additional investigation at fifteen-minute intervals between stages 16 and 18½ would provide more detailed information on gastrulation. The results of other workers on gastrulation in mosquitoes indicate that, once the mesodermal cells invaginate, they immediately migrate laterally to form a thin layer without forming a mesodermal cord. The formation in *Aedes vexans* of a solid cord of mesodermal cells which later flattens does confirm the observations of Kowalevsky (1886) in *Muscidae*, Gasparini (1939) in *Syrphidae*, Butt (1934) in *Sciara*, Fish (1949) in *Lucilia sericata*, and Anderson (1962b) in *Dacus tryoni*. Varying lumenal sizes may be present in these embryos, however, whereas in *Aedes vexans* a lumen never exists. Although not seen in *Aedes vexans*, mesodermal cells have been observed to possess unique cytoplasmic inclusions. Auten (1934) observed that in *Phormia regina* yolk globules become incorporated into the invaginated mesodermal cells, while Davis (1967) found that the presumptive mesoderm of the mid-ventral band consists of yolk-rich cells. Mahowald (1963) studied the distribution of mitochondria in the blastoderm of *Drosophila melanogaster* and found that there were 1.6 times as many mitochondria on the mid-ventral (presumptive mesoderm) as on the mid-dorsal region. He concluded that the reason for this might be to provide energy for the active role the mesoderm plays in embryogenesis. The same reason could be used to explain the richness of glycogen in the mesodermal cells of *Musca vicina* (Bhuiyan and Shafig, 1959).

When gastrulation in the insect embryo is discussed, it may involve the formation of two or three germ layers, depending on whether the author chooses to recognize an entodermal layer. The origin of the midgut epithelium has long been a subject of controversy among insect embryologists. For a detailed discussion of this topic, the reader is referred to Johannsen and Butt (1941). In this work the topic will be discussed only with reference to the Diptera. Among the Cyclorrhaphans the general scheme for midgut development is that the anterior and posterior ends of the presumptive mesodermal bands become separated off as the anterior and posterior midgut rudiments respectively. The rudiments are then carried into the yolk by the stomodeal and proctodeal invaginations. Each rudiment gives rise to paired ribbons which fuse in the mid-region. With their medial and lateral growth, the yolk mass comes to be enclosed by the midgut epithelium.

Noack (1901), DuBois (1932), Poulson (1937), and Guichard (1971) designates these rudiments as entoderm. Some authors have observed the midgut ribbons growing outward from the tips of the stomodeal and proctodeal invaginations and have concluded that the midgut is ectodermal in origin (Eastham, 1930). Among the Culicidae a similar type of midgut derivation has been described by Larsen (1952), Christophers (1960), Idris (1960), and Davis (1967). Ivanova-Kazas (1947) described the paired entodermal ribbons as growing outward from the bottom of the stomodeal and proctodeal tips. The growth of the ribbons resulted from intense cell division at the ends of these invaginations. Guichard (1971) was unable to trace the entodermal cells because they did not possess histological characteristics that permitted them to be distinguished from other cells.

In *Aedes vexans* the cells that form the midgut epithelium definitely do not arise from cell clusters separated off from the presumptive mesodermal band. Careful examination of serial sections of midgut cells shows that they grow outward from the tips of the stomodeal and proctodeal invaginations but arise from a migration of the mesodermal cells (derived from the inner layer) surrounding the invagination rather than from the ectodermal cells of the invagination itself. Also, there is no area of intense cellular proliferation at the point where these mesodermal cells begin to migrate outward. Since the midgut does not arise from discrete cellular masses or centers of proliferation, there seems to be no morphological justification for recognizing an entodermal layer. At this stage in *Drosophila melanogaster*, Poulson (1937) distinguished between entoderm and mesoderm on the basis of the fate of the cells, rather than on their origin, and because the midgut rudiments become separate from the mesoderm. In this study the latter criterion of Poulson's was used, rather than the fate of the cells. However, other criteria have been employed by different investigators. Eastham (1927) viewed the determination of the midgut as a physiological process; whatever cells happened to be appropriately located at the time when the entodermal structures are due to be formed will be determined as the midgut anlage. DuPorte (1960) stated that the midgut epithelium in insects is always entodermal. He believed that the germ layers are three groups of differentiated cells arranged according to a characteristic plan with a potentiality that is similar in all triploblasts. Much of this dilemma would be eliminated if the techniques of experimental embryology could be applied to the insect embryo. Through extirpation and transplantation experiments the potentialities of the cells that normally give rise to the midgut could be determined. If at the time of gastrulation they were capable of only forming midgut epithelium, they could then be classified as entodermal cells.

In *Aedes vexans* the development of the ventral nerve cord begins at stage 26 with the separation of the neurogenic cells from the ectoderm. These cells proliferate and form a mass on each side of the midline; they eventually fuse medially. Each of these paired masses is a ganglion, and later axons interconnect them with each other and with other ganglia. None of the thoracic and abdominal ganglia fuse prior to hatching. Most works on dipteran embryology have provided little detail on the development of the nervous system. The description of nerve cord development in *Aedes vexans* supports the findings of most other investigators in this area (Schaefer, 1938; Anderson, 1962b). The one divergent point is that a medial cord, described in other dipterans, was not seen in the ventral nerve cord of *Aedes vexans*. This median cord has been described by Poulson (1937) and Schaefer (1938). It arises from the proliferation of ectodermal cells along the invaginated bottom of the neural groove.

The brain begins development at stage 46 from two paired proliferative areas in the cephalic segment. These areas do not correspond to the segments of the brain. Toward the end of embryogeny the brain becomes demarcated into a protocerebrum, deutocerebrum, and tritocerebrum. The corpora cardiaca and allata are budded off from the ectodermal wall of the cephalic segment and fuse to the protocerebrum. The frontal ganglion of the stomodeal nervous system develops from the anterior surface of the pharynx. Schaefer (1938) working on *Phormia regina* and Larsen (1952) on *Culiseta inornata* were unable to distinguish brain segmentation before hatching. This was clearly observed in *Aedes vexans*. Larsen (1952) also stated that the source of the neurilemma is unknown. In the present study it was shown that this structure develops from peripheral cells of the ganglionic mass.

FIGURES

KEY FOR FIGURES

A	anterior	mg	midgut
ag	abdominal ganglia	Mp	micropyle
Am	adductor of mandible	mr	median retractor of flabella
Ant	antenna	mt	malpighian tubule
ap	anal papillae	nc	nerve cord
bl	blastoderm	Oe	oenocytes
c	colon	p	posterior
Ca	corpus allatum	pa	pyloric ampulla
Cc	corpus cardiacum	PC	protocerebrum
cc	cerebral commissure	pc	pole cell
D	dorsal	pgc	primordial germ cell
DC	deutocerebrum	ph	pharynx
dm	depressor of maxilla	Pr	proctodeum
do	dorsal organ	pt	palatum
dvm	dorso-ventral muscle	Pv	proventriculus
E	ectoderm	rm	retractor of maxilla
e	eye	s	serosa
ec	epidermal cell	SG	subesophageal ganglion
en	endochorion	Som	somatic mesoderm
es	esophagus	SpG	supraesophageal ganglion
fg	frontal ganglion	Spm	splanchnic mesoderm
gc	gastric caeca	St	stomodeum
H	heart	t	tracheoles
hg	hypocerebral ganglion	TC	tritocerebrum
hs	hatching spine	tg	thoracic ganglia
ic	intra-ganglionic connective	tr	trachea
il	ileum	Tt	tracheal trunk
itc	inter-ganglionic connective	V	ventral
Lbr	labrum	vm	vitelline membrane
lr	lateral retractor of flabella	Y	yolk

Fɪɢ. 1. Egg at time of oviposition.

Fɪɢ. 2. Longitudinal section through the posterior pole of an egg at stage 1. Arrows indicate the boundaries of the convex disc of polar granules. 1196×

Fɪɢ. 3. Longitudinal section through the posterior pole of an egg at stage 2, showing the scattered polar granules. 792×

Fɪɢ. 4. Longitudinal section through the micropyle of an egg at stage ½. 712×

Fɪɢ. 5. Section through the head of a sperm found in a stage 0 egg. 1424×

Fɪɢ. 6. Meiosis I in a stage ½ egg. 1140×

Fɪɢ. 7. Longitudinal section through the mid-region of an egg at stage 1, showing the two polar bodies at the egg surface. 944×

Fɪɢ. 8. Fusion of the male and female pronuclei at stage 2. 800×

Fɪɢ. 9. Section through a cleavage nucleus at stage 4. 760×

Fɪɢ. 10. Longitudinal section through a segment of the blastoderm in a stage 9 embryo. 333×

Fɪɢ. 11. Cross section through the blastoderm of a stage 12 embryo. Note the asynchrony between the dividing cells of the dorsal and dorso-lateral blastoderm, and those of the lateral and ventral blastoderm. 280×

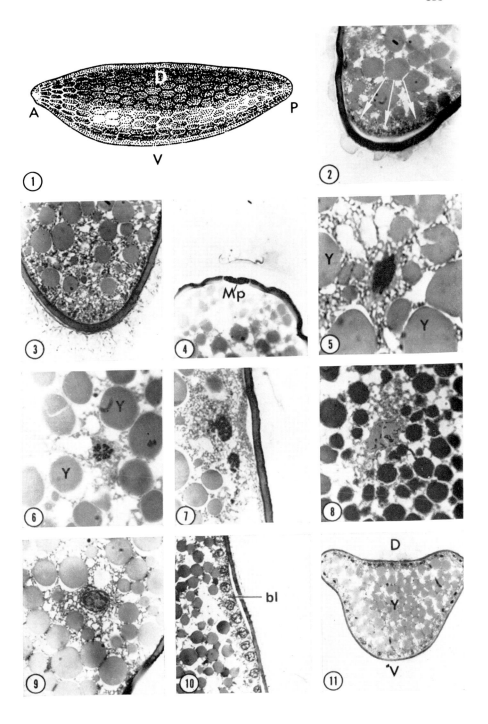

FIG. 12. Longitudinal section through the posterior pole of an egg at stage 5, showing the differentiation of a pole cell. Arrows indicate boundary of the latter. 640×

FIG. 13. Longitudinal section through the posterior blastoderm of a stage 7 egg, showing three pole cells outside the blastoderm. 568×

FIG. 14. Longitudinal section through the posterior pole of a stage 13 blastoderm, showing the inward migration of the pole cells. Arrow indicates large granule resulting from coalescence of several smaller ones. 627×

FIG. 15. Longitudinal section through an egg at stage 6. Arrows indicate yolk cells. 265×

FIG. 16. Section through the blastoderm of a stage 9 embryo. Arrow indicates a yolk cell which has migrated into the blastoderm. 568×

FIG. 17. Section through a yolk cell in a stage 13 embryo. 1280×

FIG. 18. Cross section through a stage 16 embryo. Note the dorsal columnar cells which will give rise to the germ band. 296×

FIG. 19. Cross section through the posterior pole of a stage 16 embryo, showing the point of invagination of the pole cells. 548×

FIG. 20. Longitudinal section through an embryo at stage 18½, showing the elongated germ band and the terminal position of the pole cells. 320×

FIG. 21. Cross section through an embryo at stage 18½, showing the medial movement of the amniotic membrane. Arrow indicates amniotic cavity. 284×

FIG. 22. Same as Fig. 21, but the amnion has fused mid-dorsally. 457×

FIG. 23. Cross section through an embryo at stage 18½. Note cord-like configuration of mesodermal cells. 195×

183

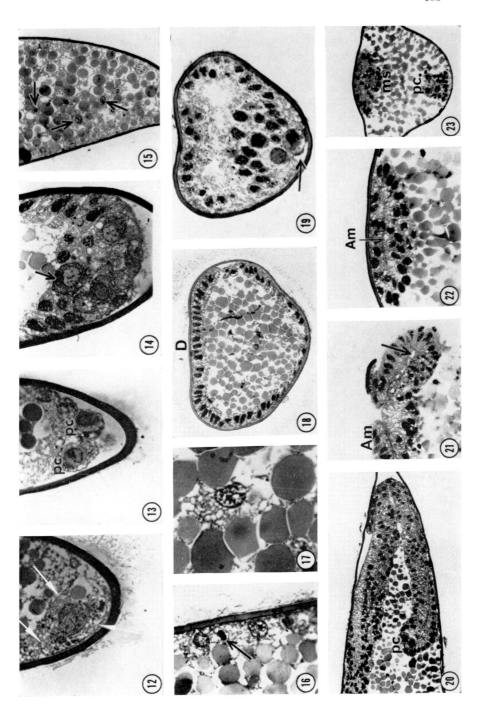

FIG. 24. A-C represent lateral views of embryos at stage 21, 31, and 56 respectively; D is a ventral view of a stage 56 embryo.

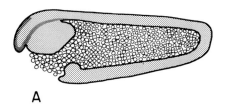

A

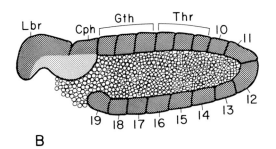

B

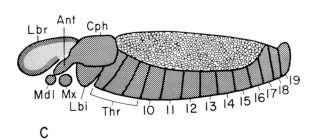

C

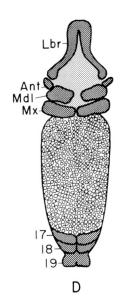

D

Abbreviations

Ant	antenna
Cph	cephalon
Gth	gnathal
Lbi	labium
Lbr	labrum
Mdl	mandible
Mx	maxilla
Thr	thorax

186

FIG. 25. Longitudinal section through the germ band of a stage 21 embryo, showing the ectodermal and segmented mesodermal layers. 312×

FIG. 26. Longitudinal section through the dorsal portion of a stage 71 embryo. Note the ectodermal invaginations demarcating the intersegmental regions; arrows indicate tracheal rudiments. 440×

FIG. 27. Cross section through the anterior labral segment of a stage 21 embryo. Note the bilobed nature of this segment. 285×

FIG. 28. Cross section through the labral segment of a stage 56 embryo. Arrows indicate the developing feeding brushes. 235×

FIG. 29. Cross section through the labrum of a stage 96 embryo. Note that the feeding brushes are now projecting ventrally and that the dorsal epidermis is undulated. 332×

FIG. 30. Cross section through the antennal rudiment of a stage 31 embryo. 460×

FIG. 31. Cross section through the area of antennal connection in a stage 86 embryo. 268×

FIG. 32. Cross section through the mandibular region of a stage 56 embryo. Note that the antennae and maxillae are also present in this section. 268×

FIG. 33. Longitudinal section through the mouthparts of a stage 56 embryo. Arrows indicate the trisegmented nature of the subesophageal ganglion. 408×

FIG. 34. Cross section through the abdominal region of a stage 66 embryo, showing the dorsal organ. 268×

FIG. 35. Cross section through a stage 56 embryo, showing the vitelline membrane adherent to the serosa and below the endochorion. 625×

FIG. 36. Longitudinal section through a stage 31 embryo, showing the stomodeal and proctodeal invaginations. 192×

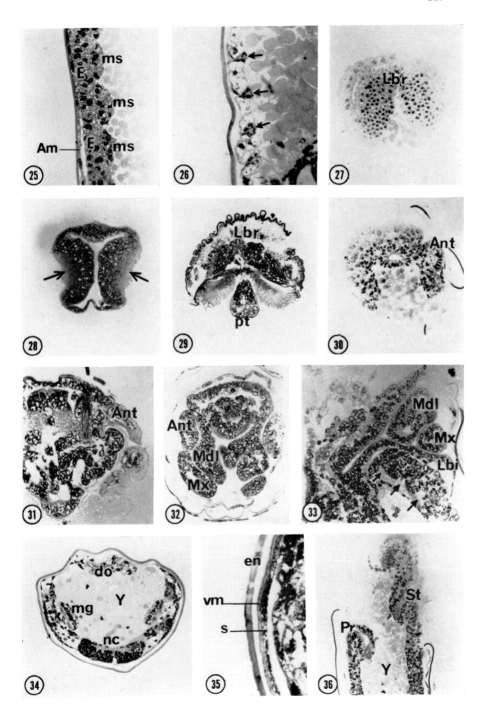

Fig. 37. Cross section through the antennal region of a stage 31 embryo. Arrow indicates mesodermal cells surrounding the stomodeum. 528×

Fig. 38. Longitudinal section through a stage 51 embryo, showing the dorsal pharyngeal invagination of the esophagus. 256×

Fig. 39. Longitudinal section through a stage 81 embryo, showing the layers of the proventriculus. 400×

Fig. 40. Cross section through the proventriculus of a stage 91 embryo. Arrows indicate the cardia. 992×

Fig. 41. Longitudinal section through a stage 46 embryo, showing the position of the proctodeum after germ band contraction. Arrows indicate mesodermal cells surrounding this structure. 268×

Fig. 42. Cross section through the proctodeum of a stage 31 embryo. Arrows indicate mesodermal cells surrounding the proctodeum. 468×

Fig. 43. Cross section through the pyloric ampulla of a stage 96 embryo. 400×

Fig. 44. Cross section through the ileum of a stage 96 embryo. Note the thin surrounding mesodermal layer indicated by arrows. 568×

Fig. 45. Cross section through the colon of a stage 96 embryo. Note the enclosing mesodermal layer. 880×

Fig. 46. Cross section through the posterior portion of the eighth abdominal segment in a stage 66 embryo. Note the presence of four pole cells in the anal lobe. 440×

Fig. 47. Cross section through the anal papillae of a stage 86 embryo. 368×

Fig. 48. Longitudinal section through the posterior region of a stage 46 embryo, showing the development of the malpighian tubules. 144×

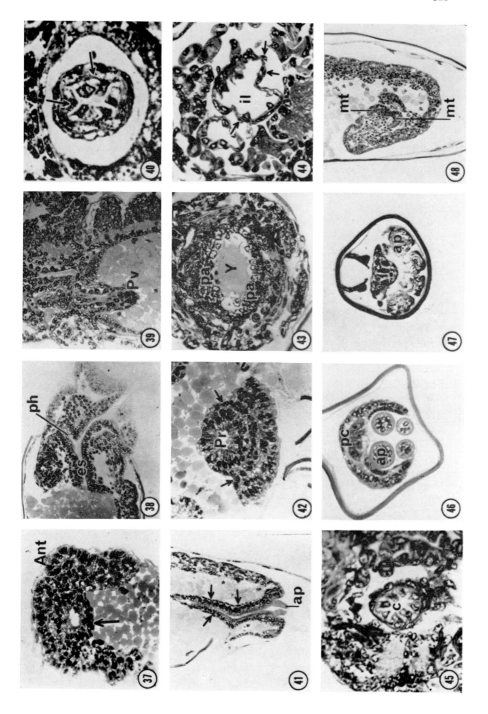

FIG. 49. Cross section through the malpighian tubules in a stage 46 embryo. 536×

FIG. 50. Cross section through the posterior tip of the stomodeum in a stage 51 embryo, showing the mesodermal cells migrating posteriorly to form the midgut. 616×

FIG. 51. Cross section through the tip of the proctodeum in a stage 51 embryo, showing the mesodermal cells migrating anteriorly to form the midgut. 428×

FIG. 52. Cross section through the abdominal region of a stage 51 embryo, showing the position of the two cords of midgut cells. 236×

FIG. 53. Cross section through a cord of midgut cells in a stage 51 embryo. Note the separation of splanchnic and somatic mesoderm. 848×

FIG. 54. Longitudinal section through a stage 96 embryo, showing gastric caeca and the fused, unsegmented subesophageal ganglion. 192×

FIG. 55. Cross section through a portion of the midgut in a stage 96 embryo. Arrows indicate splanchnic mesodermal cells. 1136×

FIG. 56. Cross section through the cephalic segment of a stage 46 embryo. Arrows indicate regions of proliferation of brain cells. 344×

FIG. 57. Frontal section through the cephalic region of a stage 76 embryo, showing the proto- and deutocerebrum. 212×

FIG. 58. Cross section through the supraesophageal ganglion of a stage 61 embryo. 232×

FIG. 59. Cross section through the deuto- and tritocerebrum of a stage 61 embryo. 224×

FIG. 60. Cross section through the supraesophageal ganglion in a stage 56 embryo, showing the differentiation of the cerebral commissure. 400×

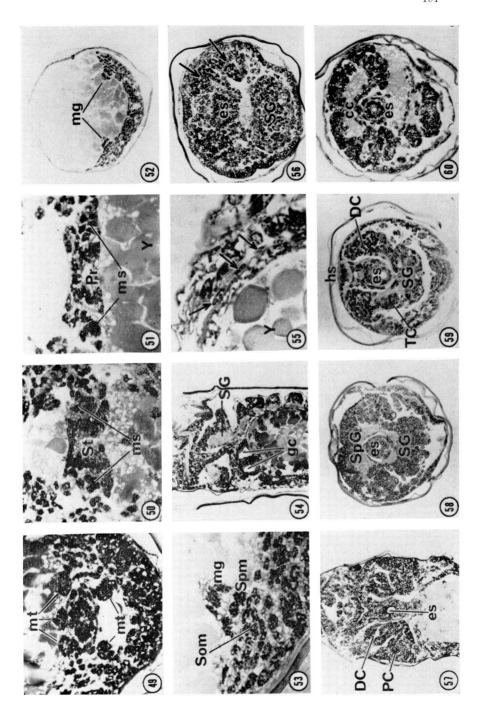

Fig. 61. Section through the brain of a stage 96 embryo. Arrows indicate the portion of the thin surrounding membrane. 1080×

Fig. 62. Section through the brain of a stage 66 embryo, showing some dividing neural cells. 912×

Fig. 63. Cross section through the frontal ganglion of a stage 91 embryo. 344×

Fig. 64. Cross section through the posterior portion of the brain in a stage 91 embryo, showing the hypocerebral ganglion. 316×

Fig. 65. Cross section through the mid-region of the cephalic segment in a stage 71 embryo, showing the development of the corpus cardiacum and corpus allatum. 332×

Fig. 66. Cross section through the corpus cardiacum and protocerebrum in a stage 71 embryo. 467×

Fig. 67. Cross section through the developing subesophageal ganglion in a stage 46 embryo. Arrows indicate the lobes of this ganglion. 180×

Fig. 68. Cross section through the subesophageal ganglion of a stage 56 embryo, showing the differentiation of the intraganglionic connective. 568×

Fig. 69. Frontal section through the nerve cord of a stage 96 embryo, showing the inter- and intraganglionic connectives. 296×

Fig. 70. Cross section through a stage 31 embryo; arrow shows the division of neuroblasts. Note that the ventral extension of the germ band is also shown here. 480×

Fig. 71. Longitudinal section through the nerve cord of a stage 56 embryo. Arrow indicates the developing neuropile. 400×

Fig. 72. Longitudinal section through a stage 96 embryo, showing all the ganglia of the central nervous system. 148×

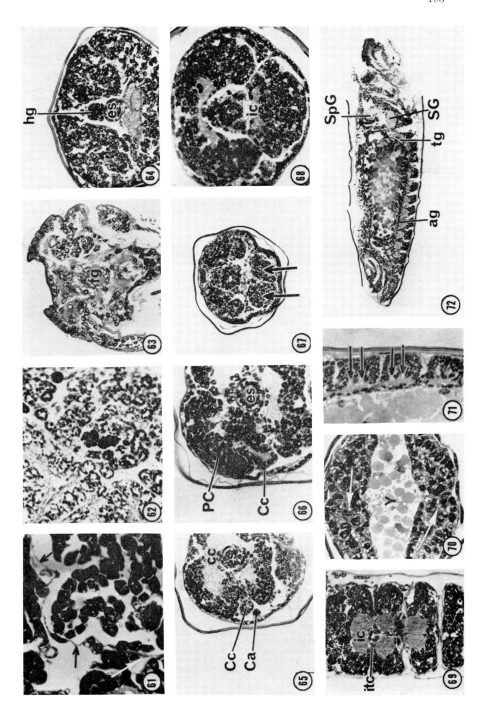

Fɪɢ. 73. Cross section through one developing eye in a stage 86 embryo. 368×

Fɪɢ. 74. Section through one developing eye in a stage 91 embryo. Note the presence of numerous pigment granules. 752×

Fɪɢ. 75. Cross section through the labrum in a stage 96 embryo, showing the undulated cuticle and position of the epidermal cells. 400×

Fɪɢ. 76. Cross section through a stage 51 embryo, showing the initial differentiation of the hatching spine. 440×

Fɪɢ. 77. Cross section of a stage 91 embryo, showing the hatching spine in longitudinal section. 308×

Fɪɢ. 78. Cross section through an abdominal segment of a stage 96 embryo, showing the location of the oenocytes. 466×

Fɪɢ. 79. Cross section through the eighth abdominal segment of a stage 61 embryo. Arrows indicate ectodermal invaginations that will develop into the tracheal trunks of the respiratory siphon. 536×

Fɪɢ. 80. Longitudinal section through the posterior portion of a stage 81 embryo. Arrows indicate the S-shaped bend in the tracheal trunks of the respiratory siphon. 308×

Fɪɢ. 81. Cross section through the respiratory siphon of a stage 96 embryo, showing the developing tracheoles in relation to one of the tracheal trunks. 856×

Fɪɢ. 82. Cross section through a tracheal tube in an abdominal segment of a stage 91 embryo. 488×

Fɪɢ. 83. Cross section through a longitudinal tracheal trunk in an abdominal segment of a stage 91 embryo. 400×

Fɪɢ. 84. Cross section through the labrum of a stage 46 embryo, showing the mesodermal cells within this segment. 364×

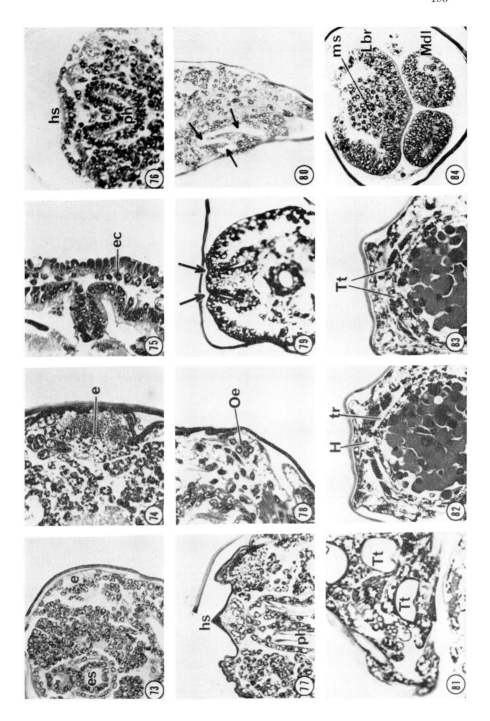

Fɪɢ. 85. Cross section through the labrum of a stage 56 embryo, showing the median and lateral retractors of the flabella. 344×

Fɪɢ. 86. Cross section through the labrum of a stage 96 embryo, showing the median and laberal retractors of the flabella. Arrow indicates point of attachment of retractors. 416×

Fɪɢ. 87. Cross section through the mandibular region of a stage 96 embryo, showing the adductors of the mandible. 344×

Fɪɢ. 88. Cross section through the mandibular region of a stage 91 embryo. Arrow indicates the point of insertion and a part of the abductors of the mandible. 432×

Fɪɢ. 89. Cross section through the maxillary region of a stage 96 embryo, showing the depressors and retractors of the maxillae. 384×

Fɪɢ. 90. Frontal section through a stage 96 embryo, showing the oblique longitudinal muscles of the abdomen. 256×

Fɪɢ. 91. Same as Fig. 90 but showing the antero-posteriorly directed longitudinal muscles of the abdomen. 344×

Fɪɢ. 92. Cross section through an abdominal segment of a stage 96 embryo, showing a muscle of the dorso-ventral series. 476×

Fɪɢ. 93. Cross section through the heart of a stage 96 embryo.

Fɪɢ. 94. Cross section through an abdominal segment in a stage 76 embryo. Arrows indicate location of the fat body. 368×

Fɪɢ. 95. Longitudinal section through the posterior portion of a stage 46 embryo completing germ band contraction. Note the position of pole cells. 236×

Fɪɢ. 96. Cross section through the sixth abdominal segment of a stage 51 embryo, showing the primordial germ cells in relation to the midgut and mesodermal rudiment. 200×

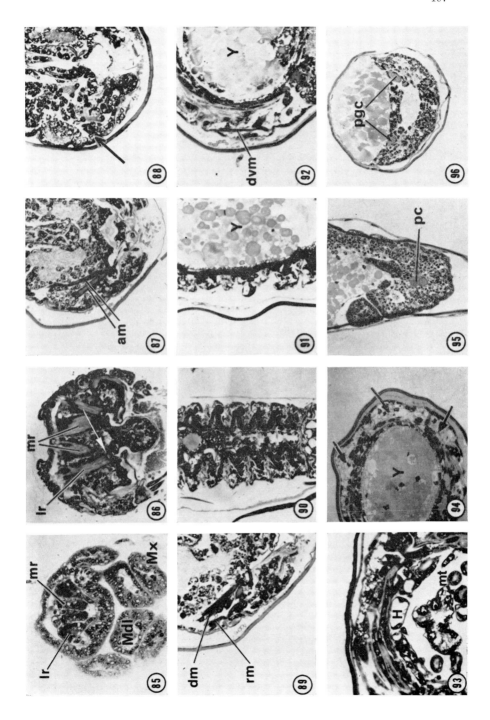

SUMMARY

At oviposition the egg is composed of a yolk matrix with an interconnecting cytoplasmic network; a distinct periplasm is absent. The small yolk spheres are located at the periphery of the egg, and the larger ones are found in the interior. Polyspermy occurs frequently.

The female pronucleus divides meiotically to produce a single polar body. The latter undergoes an additional mitotic division. The two polar bodies degenerate at the egg surface. The male and female pronuclei fuse in the anterior portion of the egg, and shortly afterwards nuclear cleavage begins. Between stages 6 and 7 the cleavage nuclei migrate to the egg surface to form a syncytial blastoderm. At stage 14 the cell boundaries are completed.

The pole cells begin to differentiate during stage 5 at the posterior egg pole. The number of pole cells is about eleven; these never undergo mitosis until just before hatching. The cytoplasm of these cells contains polar granules and small yolk spheres.

Primary yolk cells differentiate from those cells which do not migrate to the surface at blastoderm formation. These cells are usually ameboid and do not possess a cell boundary. Secondary yolk cells arise from blastoderm cells which reenter the yolk matrix.

Between stages 14 and 15 the cells of the dorsal blastoderm become columnar and give rise to the germ band. The lateral edges of the germ band develop into the amnion, while the cells of the lateral and ventral blastoderm give rise to the serosa. The vitelline membrane is secreted by the serosa between stages 36 and 46.

The prospective mesodermal cells lie along the mid-line region of the germ band and at gastrulation migrate ventrally.

Segmentation of the germ band begins at stage 21. Ultimately nineteen segments are formed. The pregnathal segments are the labrum, antennae, and cephalon. These are followed by the mandibles, maxillae, labium, three thoracic and ten abdominal segments. Segments 17, 18, and 19 fuse to become the eighth abdominal segment. The three gnathal segments migrate anteriorly to underlie the cephalic region. Blastokinesis occurs between stages 36 and 46.

The stomodeal and proctodeal invaginations are ectodermal in origin and occur at stage 26. The former gives rise to the foregut and its specializations such as the pharynx, salivary glands, and proventriculus. The proctodeum differentiates into the hindgut, malpighian tubules, and anal papillae. The midgut is derived from an outward migration of the mesodermal cells that surround the stomodeal and proctodeal invaginations. Eight gastric caeca develop as outpouchings of the anterior border of the midgut.

The brain arises from localized proliferations of ectoderm in the dorso-lateral and mid-lateral regions of each side of the cephalic segment. The frontal ganglion buds off from the anterior surface of the pharynx and grows posteriorly along the dorsal surface of the esophagus; at the level of the circumesophageal nerve it enlarges to form the hypocerebral ganglion. The corpora allata and corpora cardiaca are budded off from the cephalic walls and fuse to the postero-lateral margin of the protocerebrum. The ventral ganglia arise from proliferations of the ectoderm on each side of the midline. The ganglia of the gnathal segments become localized in the labium and eventually fuse to form the subesophageal ganglion. This structure is connected anteriorly to the brain by the tritocerebral lobes. The thoracic and abdominal ganglia remain unfused at the time of hatching.

The rudiments of the respiratory siphon arise as a pair of ectodermal invaginations at the base of the respiratory siphon. The longitudinal and circular trachea differentiate from ectodermal cells that invaginate at the dorsal and lateral intersegmental areas.

The medial and lateral retractors are the first and largest muscles to differentiate. They are derived from mesodermal cells found within the labrum. The muscles of the appendicular structures are derived from mesodermal cells located within the cavity of these structures. The musculature of the thoracic and first seven abdominal segments is derived from the somatic mesoderm; the heart and fat body also differentiate from this rudiment. A thin layer of splanchnic mesoderm encloses the midgut but apparently does not differentiate into a muscular layer.

At the time of germ band contraction the pole cells divide themselves into three groups. Two of these form the gonads, while the third migrates to the postero-dorsal area of the anal lobe and remains there until the time of hatching.

REFERENCES CITED

Abou-Aly, A. M. 1968. Bionomics of *Psorophora varipes* (Coquillett) (Diptera: Culicidae). Ph.D. dissertation, University of Illinois. 57 pp.

Agrell, I. 1963. Mitotic gradients in the early insect embryo. *Arkiv für Zoologi* 15:143-148.

Anderson, D. T. 1961. A differentiation centre in the embryo of *Dacus tryoni*. *Nature* 190:560-561.

———. 1962a. The epigenetics of the larva in Diptera. *Acta Zoologia* 43:221-228.

———. 1962b. The embryology of *Dacus tryoni* (Frogg) (Diptera, Trypetidae) (= Tephritidae), the Queensland fruit fly. *Journal of Embryology and Experimental Morphology* 10:248-292.

———. 1963. The embryology of *Dacus tryoni*. 2. Development of imaginal discs in the embryo. *Journal of Embryology and Experimental Morphology* 11:339-351.

———. 1964. The embryology of *Dacus tryoni*. 3. Origins of imaginal rudiments other than the principal discs. *Journal of Embryology and Experimental Morphology* 12:65-75.

Auten, M. 1934. The early embryological development of *Phormia regina:* Diptera (Calliphoridae). *Annals of the Entomological Society of America* 27:481-499.

Bantock, C. 1961. Chromosome elimination in *Cecidomyidae*. *Nature* 190:466-467.

Berezin, V. V., M. P. Chumakov, V. N. Bashkirtsev, and B. F. Semenov. 1971. Arboviral infections in the Volga delta. Translation of #510 of Naval Medical Research Unit #3, Cairo, Egypt, c/o Spanish Embassy. 3 pp. From Akademiya Nauk SSSR [Meditsina] Oct. 19-21 (pt. 2):137-138.

Bhaskaran, G., V. Ramakrishnan, and C. Adeesan. 1970. Effects of Benzamide on embryonic development of the housefly. *Development, Growth and Differentiation* 11:265-276.

Bhuiyan, N. I., and S. A. Shafig. 1959. The differentiation of the posterior pole plasm in the housefly, *Musca vicinia* Macquart. *Experimental Cell Research* 16:427-429.

Butt, F. H. 1934. Embryology of *Sciara* (Sciaridae: Diptera). *Annals of the Entomological Society of America* 27:565-579.

———. 1957. The role of the premandibular or intercalary segment in head segmentation of insects and other arthropods. *Transactions of the American Entomological Society* 83:1-30.

Christophers, S. R. 1960. *Aedes aegypti (L.) the yellow fever mosquito; its life history, bionomics and structure.* Cambridge: Cambridge University Press. 738 pp.

Counce, S. J. 1961. The analysis of insect embryogenesis. *Annual Review of Entomology* 6:295-312.

———. 1963. Developmental morphology of polar granules in Drosophila. Including observations on pole cell behavior and distribution during embryogenesis. *Journal of Morphology* 112:129-145.

Davis, C. W. C. 1967. A comparative study of larval embryogenesis in the mosquito *Culex fatigans* Wiedmann (Diptera: Culicidae) and the sheep-fly *Lucilia sericata* Meigen (Diptera: Calliphoridae). *Australian Journal of Zoology* 15:547-579.

DeCoursey, J. D., and A. P. Webster. 1952. A method of clearing the chorion of *Aedes sollicitans* (Walker) eggs and preliminary observations on their embryonic development. *Annals of the Entomological Society of America* 45:625-632.

Demerec, M. 1950. *Biology of Drosophila.* New York: John Wiley. 632 pp.

DuBois, A. M. 1932. A contribution to the embryology of Sciara (Diptera). *Journal of Morphology* 54:161-192.

DuPorte, E. M. 1960. Gastrulation and the entoderm problem in insects. *Annals of the Entomological Society of Quebec* 6:45-52.

———. 1963. The postantennal region of the insect embryo. *Canadian Journal of Zoology* 41:909-912.

Eastham, L. 1927. A contribution to the embryology of *Pieris rapae. Quarterly Journal of Microscopical Science* 71:353-394.

———. 1930. The formation of germ layers in insects. *Biological Review* 5:1-29.

Ede, D. A. 1956. Studies on the effects of some genetic lethal factors on the embryonic development of *Drosophila melanogaster.* I. A preliminary survey of some sex linked lethal stocks and an analysis of the mutant Lff11. *Archiv für Entwicklungsmechanik der Organismen* 148:416-436.

————, and S. J. Counce. 1956. A cinematographic study of the embryology of *Drosophila melanogaster*. *Archiv für Entwicklungsmechanik der Organismen* 148:259-266.

Escherich, K. 1901. Das Insekton-Entoderm. *Biologisches Zentralblatt* 21: 416-431.

Fish, W. A. 1947a. Embryology of *Lucilia sericata* Meigen (Diptera: Calliphoridae). Pt. II. The blastoderm yolk cells and germ cells. *Annals of the Entomological Society of America* 40:677-687.

————. 1947b. Embryology of *Lucilia sericata* Meigen (Diptera: Calliphoridae). Pt. I. Cell cleavage and early embryonic development. *Annals of the Entomological Society of America* 40:15-28.

————. 1949. Embryology of *Phaenicia sericata* (Meigen) (Diptera: Calliphoridae). Pt. III. The gastrular tube. *Annals of the Entomological Society of America* 42:121-133.

————. 1952. Embryology of *Phaenicia sericata* (Meigen) (Diptera: Calliphoridae). Pt. IV. The inner layer and mesenteron rudiments. *Annals of the Entomological Society of America* 45:1-22.

Gambrell, F. L. 1933. The embryology of the black fly. *Simulium pictipes* (Hagen). *Annals of the Entomological Society of America* 26:641-671.

Gasparini, F. 1939. Notes on the embryological development of *Eristalis tenax* (Linn.). Syrphidae. *Wesmann Collector, San Francisco* 3:38-43.

Guichard, Marcelle. 1971. Étude in vivo du développement embryonnaire de *Culex pipiens*. Comparaison avec *Calliphora erythrocephala* (Diptera). *Annales de la Société entomologique de France* (N.S.), 7:325-341.

Hallez, P. 1886. Loi de l'orientation de l'embryon chez les insectes. *Compte rendu hebdomadaire des séances de l'Académie des sciences* 103:606-608.

Harber, P. A. 1969. Morphological and cytological observations on the early development of *Culiseta inornata* (Williston). Ph.D. dissertation, Iowa State University. 75 pp.

————, and J. Mutchmor. 1970. The early embryonic development of *Culiseta inornata* (Diptera: Culicidae). *Annals of the Entomological Society of America* 63:1609-1614.

Hasper, M. 1911. Zur Entwicklung der Geschlechtsorgane von *Chironomus*. *Zoologischer Jahrbücher* (Anat.), 31:543-612.

Huettner, A. F. 1940. Differentiation of the gonads in the embryo of *Drosophila melanogaster*. *Genetics* 25:121.

Idris, B. E. M. 1960. Die Entwicklung im normalen Ei von *Culex pipiens* L. (Diptera). *Zeitschrift für Morphologie und Ökologie der tiere* 49:387-429.

Ivanova-Kazas, O. M. 1947. Stages of the embryonic development of Anopheles maculipennis. *C. R. Academy of Science, U.R.S.S. Moscow* 56:325-327.

————. 1964. Embryonic development in *Heteropeza pygmaea* and *Miastor*. *Vestnik Leningrad Skogo Universiteta Seriia Biologii* 21:12-27.

Johannsen, O. A., and F. H. Butt. 1941. *Embryology of insects and myriopods*. McGraw-Hill Book Co., New York.

Korschelt, E., and K. Heider. 1899. *Textbook of the embryology of invertebrates.* Vol. III, 441. New York: Macmillan.

Kowalevsky, A. 1886. Zur embryonalen Entwicklung der Musciden. *Biologisches Zentralblatt* 6:49-54.

Kumé, Matazo, and Katsuma, Dan. 1968. *Invertebrate embryology.* Washington: NOLIT. 605 pp.

Larsen, J. R. 1952. A study of the pre-larval development of the mosquito *Culiseta inornata* (Williston). Master's thesis, University of Utah. 61 pp.

Mahowald, A. P. 1962. Fine structure of pole cells and polar granules in *Drosophila melanogaster. Journal of Experimental Zoology* 151:201-216.

————. 1963. Ultrastructural differentiations during formation of the blastoderm in the *Drosophila melanogaster* embryo. *Developmental Biology* 8: 186-204.

Nicholson, A. J. 1921. The development of the ovary and ovarian egg of a mosquito, *Anopheles maculipennis. Quarterly Journal of Microscopical Science* 65:395-448.

Noack, W. 1901. Beträge zur Entwicklungsgeschichte der Musciden. *Zeitschrift für wissenschaftliche Zoologie* 70:1-57.

Poulson, D. F. 1937. The embryonic development of *Drosophila melanogaster.* Paris: Herman & Cie. 56 pp.

————. 1947. The pole cells of Diptera, their fate and significance. *Proceedings of the National Academy of Sciences* 33:182-184.

————. 1950. Histogenesis, organogenesis and differentiation in the embryo of *Drosophila melanogaster* Meigen. In *Biology of drosophila.* M. Demerec, ed. New York: John Wiley. Pp. 168-274.

————, and D. F. Waterhouse. 1960. Experimental studies on pole cells and mid-gut differentiation in Diptera. *Australian Journal of Biological Science* 13:541-567.

Rabinowitz, M. 1941. Studies on the cytology and early embryology of the egg of *Drosophila melanogaster. Journal of Morphology* 69:1-49.

Rosay, B. 1959. Gross external morphology of embryos of *Culex tarsalis* (Coquillet). Diptera: Culicidae. *Annals of the Entomological Society of America* 52:481-484.

Schaefer, P. E. 1938. The embryology of the central nervous system of *Phormia regina* Meigen. *Annals of the Entomological Society of America* 31:92-111.

Snodgrass, R. E. 1935. *Principles of insect morphology.* New York: McGraw-Hill. 667 pp.

————. 1960. Facts and theories concerning the insect head. *Smithsonian Miscellaneous Collections* 142:1-61.

Sonnenblick, B. P. 1950. Early embryology of *Drosophila melanogaster.* In *Biology of Drosophila.* M. Demerec, ed. New York: John Wiley. Pp. 62-167.

Telford, A. D. 1957. The pasture *Aedes* of Central and Northern California. The egg stage: gross embryology and resistance to desiccation. *Annals of the Entomological Society of America* 50:537-543.

Trpis, M. 1970. A new bleaching and decalcifying method for general use in zoology. *Canadian Journal of Zoology* 48:892-893.

Weismann, A. 1863. Die Entwickelung der Dipteren im Ei. *Zeitschrift für wissenschaftliche Zoologie* 13:107-220.

West, Judith A., George E. Cantwell, and T. J. Shortino. 1968. Embryology of the housefly, *Musca domestica* (Diptera: Muscidae), to the blastoderm stage. *Annals of the Entomological Society of America* 61:13-17.

INDEX

Adult: aggregation, 55, 56, 57, 62, 69, 71-73, 89-90, 95, 113-117; antagonist, 91; appearance, 54-55, 111, 112; associated species, 56, 62-64, 67; colonization, 95-99; copulation, 80, 82, 85, 90, 97-99; dispersal, 64-78, 80, 89; emergence, 65; excretion, 88; feeding, 55, 57, 66, 78-81; focal distribution, 55-57; gynandromorph, 55; host, 78-81, 92, 96; latency, 90; longevity, 90; maturation, 83-85; mortality factors, 57, 95; oviposition, 85-88, 96; resistance, 95; seasonal occurrence, 56; secretion, 88-89; toxinosis, 94-95; trapping, 57-64, 67-70, 79, 85, 89, 116

Aedes: aegypti, 12, 138, 139; *atlanticus*, 27, 64; *atropalpus*, 37, 38; *aurifer*, 48; *barri*, 46, 48; *campestris*, 48; *canadensis*, 15, 16, 26, 27, 28, 48, 62, 63; *cinereus*, 26, 28, 46, 48, 55; *communis*, 28, 48, 64; *dianteus*, 48; *dorsalis*, 26, 48; *dupreei*, 15, 16, 26, 28, 48; *excrucians*, 46, 48; *fitchii*, 48; *flavescens*,

48; *grossbecki*, 48; *intrudens*, 48; *mariae*, 37; *nigromaculis*, 48; *punctor*, 27, 48, 64; *sollicitans*, 26, 28, 64, 65, 66, 74; *sticticus*, 15, 16, 26, 27, 28, 48, 62, 63, 64; *stimulans*, 15, 16, 26, 27, 28, 37, 48, 55, 63, 64, 79, 80, 81; *taeniorhynchus*, 66, 71, 80; *thibaulti*, 26; *trichurus*, 28; *triseriatus*, 63, 64, 67; *trivittatus*, 15, 16, 26, 27, 28, 48, 61, 62, 63, 64, 80, 81; *vexans*, 6; *vexans nipponii*, 6, 7, 93, 94; *vexans nocturnus*, 6, 7, 53, 78

Agamomermis, 53, 91

Alfalfa, 56, 66

Algae, 46

Alimentary canal, 152

Alphavirus spp., 93

Amnion, 145, 147, 148, 151, 200; differentiation from serosa cells, 145; dorsal growth, 151. *See also* Embryonic membranes

Amniotic cavity, 148

Anal papillae, 154, 164, 200; rudiments, 154

Anopheles: barberi, 63, 64; *crucians*,

207

About the Authors

WILLIAM R. HORSFALL has been professor of entomology at the University of Illinois, Urbana, for twenty-five years. He is the author of more than fifty papers on the bionomics of mosquitoes, and he has served as a consultant on mosquito control for the World Health Organization, Tennessee Valley Authority, and mosquito abatement districts.

HARLAND W. FOWLER is a lieutenant colonel, U.S. Army, working in the Office of the Surgeon General. He completed his doctoral work at the University of Illinois, Urbana, in 1968.

LOUIS J. MORETTI is presently studying at the University of Illinois Medical School, Chicago. He received his Ph.D. from the University of Illinois, Urbana, in 1971.

JOSEPH R. LARSEN is professor and head of the department of entomology as well as acting director of the School of Life Sciences at the University of Illinois, Urbana.

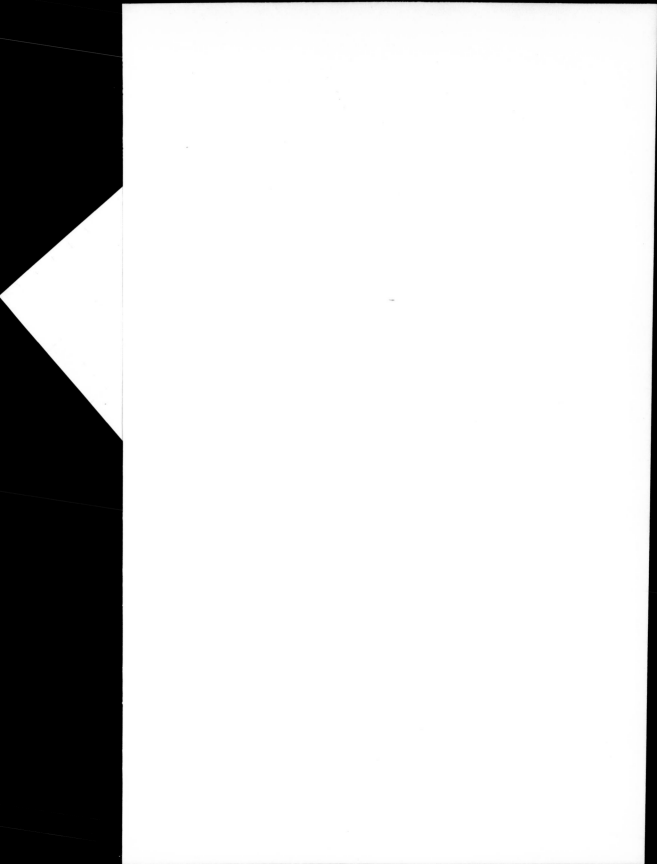

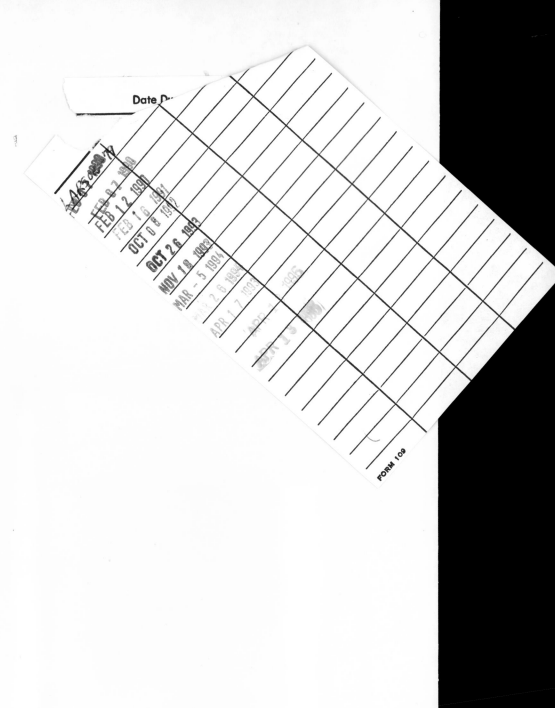